1993

Current
Communications
IN MOLECULAR BIOLOGY
Cold Spring Harbor Laboratory Press

Molecular Genetics of Early *Drosophila* and Mouse Development

Edited by

Mario R. Capecchi
Howard Hughes Medical Institute
University of Utah

 A Banbury Center Meeting

MOLECULAR GENETICS OF EARLY *DROSOPHILA* AND MOUSE DEVELOPMENT

Cover: The cover depicts what is perhaps a vulgarization of the Chinese philosophical statement of the contrasting and complementary character of yin and yang. (Cover design courtesy of Kerry Matz, Graphics Department, University of Utah; and Mario R. Capecchi, Department of Biology, University of Utah.)

All Cold Spring Harbor Laboratory Press publications may be ordered directly from Cold Spring Harbor Laboratory, Box 100, Cold Spring Harbor, New York 11724. (Phone: Continental U.S. except New York State 1-800-843-4388. All other locations [516] 367-8325.)

Titles in
Current Communications in Molecular Biology

The meeting on Molecular Genetic Analysis of Early *Drosophila* and Mouse Development was funded entirely by proceeds from the Laboratory's Corporate Sponsor Program, whose members provide core support for Cold Spring Harbor and Banbury meetings:

Alafi Capital Company
American Cyanamid Company
Amersham International plc
AMGen Inc.
Applied Biosystems, Inc.
Becton Dickinson and Company
Beecham Pharmaceuticals
Boehringer Mannheim Corporation
Ciba-Geigy Corporation/Ciba-Geigy Limited
Diagnostic Products Corporation
E.I. du Pont de Nemours & Company
Eastman Kodak Company
Genentech, Inc.
Genetics Institute
Glaxo Research Laboratories
Hoffmann-La Roche Inc.
Johnson & Johnson
Life Technologies, Inc.
Eli Lilly and Company
Millipore Corporation
Monsanto Company
Oncogene Science, Inc.
Pall Corporation
Perkin-Elmer Cetus Instruments
Pfizer Inc.
Pharmacia Inc.
Schering-Plough Corporation
Tambrands Inc.
The Upjohn Company
The Wellcome Research Laboratories,
  Burroughs Wellcome Co.
Wyeth-Ayerst Research Laboratories

Conference Participants

Rosa Beddington, Dept. of Zoology, ICRF Developmental Biology Unit, Oxford, England

Welcome Bender, Dept. of Biological Chemistry, Harvard Medical School, Boston, Massachusetts

David Bowtell, Dept. of Biochemistry, University of California, Berkeley

Mario R. Capecchi, Howard Hughes Medical Institute, University of Utah, Salt Lake City

Denis DuBoule, European Molecular Biology Laboratory, Heidelberg, Federal Republic of Germany

Peter Gruss, Dept. of Molecular Cell Biology, Max-Planck-Institute of Biophysical Chemistry, Göttingen, Federal Republic of Germany

Brigid Hogan, Dept. of Cell Biology, Vanderbilt University Medical School, Nashville, Tennessee

Alexandra Joyner, Dept. of Molecular and Developmental Biology, Mt. Sinai Hospital Research Institute, Toronto, Canada

Mark Krasnow, Dept. of Biochemistry, Stanford University Medical Center, California

Robb Krumlauf, National Institute for Medical Research, London, England

Michael Levine, Dept. of Biological Sciences, Columbia University New York, New York

Gail Martin, Dept. of Anatomy, University of California, San Francisco

Jean Marx, Science, American Association for the Advancement of Science, Washington, D.C.

William McGinnis, Dept. of Molecular Biophysics and Biochemistry, Yale University, New Haven, Connecticut

Andrew McMahon, Dept. of Cell and Developmental Biology, Roche Institute of Molecular Biology, Nutley, New Jersey

Roel Nusse, Netherlands Cancer Institute, Amsterdam

David Page, The Whitehead Institute for Medical Research, Cambridge, Massachusetts

Elizabeth Robertson, Dept. of Genetics and Development, Columbia University, New York, New York

Frank H. Ruddle, Yale University, New Haven, Connecticut

Matthew P. Scott, Dept. of Molecular, Cellular and Development Biology, University of Colorado, Boulder

Tim Stewart, Genentech, Inc., South San Francisco, California

Erwin F. Wagner, Research Institute of Molecular Pathology, Vienna, Austria

Gordon Wong, Genetics Institute, Cambridge, Massachusetts

Preface

Over the past few years, the combined molecular and genetic analysis of early *Drosophila* development has yielded a remarkable description of how the metameric pattern in the embryo is established. Following fertilization, a small number of maternally localized determinants trigger a cascade of spatially regulated transcription that successively subdivides the embryo into smaller and smaller domains. In the process of subdivision, each cell in the embryo acquires a combination of transcription factors required to specify its unique identity. As an example, in terms of specifying cell identity along the anterior-posterior axis, these transcription factors are the products of the pair-rule genes.

This volume presents a glimpse of how investigators of *Drosophila* are now attempting to push this analysis several layers deeper in order to describe molecularly how the control of temporal and spatial gene expression is accomplished. Because of the overall complexity of the problem, each investigator is developing ingenious assays to isolate a component of the system that is more amenable to in-depth analysis. The spectrum of investigators reflects the spectrum of approaches: from the analysis of transgenes in the whole organism to the analysis of in vitro transcription systems.

On the surface, early mouse development appears very different from early *Drosophila* development. Initially, development of the mouse zygote proceeds slowly with the first few cell divisions producing equivalent cells. In addition, a large proportion of early mouse development is devoted toward generating the extraembryonic tissues needed for nourishing the embryo. However, the molecular dynamics of mouse development, particularly following the onset of gastrulation, may turn out not to be that different from *Drosophila* development.

The impetus for this belief emerged from the finding that the mouse contains homologs for many of the *Drosophila* genes that control embryonic development. Furthermore, these genes are expressed in distinct temporal and spacial patterns during embryogenesis. Most recently, it has been found that for the *Antennapedia*-related homeo-box-containing genes, not only has the gene order between mouse and *Drosophila* been conserved, but the correlation that the order of genes on the chromosome reflects the order of the boundaries of gene expression along the anterior-posterior axis of the embryo has also been conserved. These observations suggest that the evolutionary conservation of these genes may not merely reflect the retention of convenient DNA-binding motifs, but rather the inheritance of a whole program for specifying positional information in the embryo.

Between April 20 and 23, 1989, investigators of *Drosophila* and mouse development met at the Banbury Center of Cold Spring Harbor Laboratory, for an intensive discussion. The purpose of this meeting was to establish a dialogue that would lead to a better appreciation of the similarities and differences in the developmental programs utilized by these two organisms. More than that, we hoped to obtain insight into the strategies and technologies employed to analyze development in these organisms and perhaps catch a glimpse of how each field may develop in the future.

The high expectations for the meeting were more than adequately met. The Banbury Center is an idyllic setting for discussion. The dialogue was lively and instructive. Clearly, the analyses of *Drosophila* and mouse development are at different stages, but both are at exciting junctures. The relative ease of genetic manipulations in *Drosophila* has yielded a wealth of informative mutant phenotypes ready for the onslaught of molecular biologists. *Drosophila* will serve as the paradigm for the analysis of development in more complex organisms, and, indeed, it has already served this purpose by providing eager molecular biologists with a route for cloning genes of potential importance for mouse development. Having identified the players, mouse biologists must now establish their roles. Hopefully, gene targeting will allow circumvention of the mouse's relatively unfavorable genetic attributes and thus provide the means for assigning a function to this interesting family of genes.

It is our hope that some of the excitement of the meeting has been translated to this small volume. This book should provide more than an update of how each participant is approaching their analysis of development of their respective organism; it should also afford an overview of how each field is evolving at this crucial point in time.

In closing, it is a pleasure for me to thank, in the name of all the participants, Jan A. Witkowski, the Director of the Banbury Center, for his energy, good humor, and administrative skills required to bring this meeting to fruition. We also thank

Nancy Ford, Managing Director of the Cold Spring Harbor Laboratory Press, and her staff, Ralph Battey and Inez Sialiano, for editing this book.

M.R.C.

Contents

Segmental Regulation of the Bithorax Complex of *Drosophila*

W. Bender,[1] J. Simon[1,] F. Karch,[2] M. O'Connor,[3] and M. Peifer[4]

[1]Department of Biological Chemistry and Molecular Pharmacology
Harvard Medical School, Boston, Massachusetts 02115

[2]Department of Animal Biology, University of Geneva
Geneva, Switzerland

[3]Department of Biochemistry and Molecular Biology
University of California at Irvine, Irvine, California 92717

[4]Biology Department, Princeton University
Princeton, New Jersey 08544

The Bithorax Complex (BX-C) and the Antennapedia Complex (ANT-C) are two clusters of genes in *Drosophila* that control the developmental fates of the different segments of the animal. Mutations in individual components of either complex transform one segmental unit into another. For example, a fly mutant for the *bithoraxoid* (*bxd*) function of BX-C has eight legs because its first abdominal segment is transformed into a thoracic segment. These transformations actually affect sections of the body called parasegments; a parasegment includes the posterior one fourth of one segment plus the anterior three fourths of the next adjacent segment.

Most of the genetic description of BX-C has been carried out by E.B. Lewis and that of ANT-C by T. Kaufman. Lewis noticed that the mutations are arranged on the chromosome in the order of the segments that they affect (Lewis 1978). Kaufman's studies of ANT-C show the same order (Kaufman et al. 1980), although ANT-C also includes several genes whose functions are not related to segmental identity.

Nine Regulatory Domains

We have constructed a molecular map of BX-C and identified the lesions of many of the mutations that cause the segmental transformations. Figure 1 shows an alignment of the DNA map with the segments of the fly most affected by the different classes of mutations. The nine different DNA regions represent

1

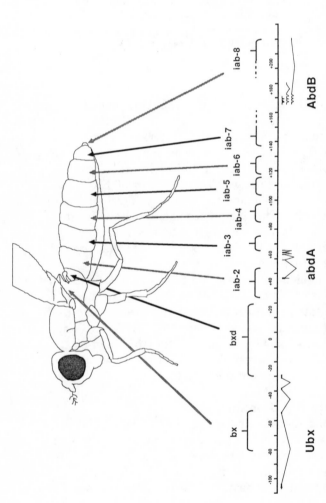

Figure 1 Alignment of BC-X with the segments of the fly. The continuous line at the bottom represents the DNA map of BC-X marked in kb. Below the DNA map are shown the three transcription units for the three homeo box protein families; all are transcribed from right to left. Above the DNA map are shown the approximate extents of the nine regulatory domains. The segment (or rather the parasegment) of the fly which is most affected by mutations in that domain (↑). (Adapted, with permission, from Peifer et al. 1987).

2

not nine different gene products but nine regulatory regions for only three gene products. The three products are called *Ultrabithorax* *(Ubx)* (with the *bx* and *bxd* regulatory regions) *abdominal-A* *(abd-A)* (with *infra-abdominal* [*iab*]-2, *iab-3*, and *iab-4* as regulatory regions), and *Abdominal-B* *(Abd-B)* (with *iab-5* to *iab-8* as regulatory regions); their transcription maps are also shown in Figure 1. Alternate splicing actually generates a family of related protein products for the *Ubx* unit (O'Connor et al. 1988; Kornfeld et al. 1989). Alternate promoters give variant protein products for *Abd-B* (Kuziora and McGinnis 1988; Celniker et al. 1989), and variant products are also suspected for *abd-A* (F. Karch et al., in prep.). The functions of the alternate products are not known, although in the case of *Ubx*, the alternate products are tissue-specific but not segment-specific (M. Gray and W. Bender, unpubl.). The protein products all include a homeo box peptide near their carboxy-terminal ends, and so all are thought to be DNA- or RNA-binding proteins.

It is not obvious why the homeo box genes lie on the chromosome in the order of the segments they affect or why the regulatory regions for each of the genes are also in order. We proposed a model invoking different chromosome architecture in different segments (Peifer et al. 1987). We suggested that in the third thoracic segment (parasegment 5), only the left-most regulatory domain (the *bx* region) is decondensed and accessible to tissue-specific transcription factors. In the next parasegment, the next regulatory domain (the *bxd* region) is also "open for business," and so on. The model explains the phenotypes of some of the dominant mutants of the complex, and it provides a molecular logic to the chromosome order, but we cannot really explain why that order would be so rigidly conserved.

Boundaries between Domains

If different areas of the chromosome are differentially decondensed, we might expect the boundaries between these regions to be critical in setting up the segment-specific architecture. Two small deletion mutants in the abdominal region give dominant phenotypes that we interpret as loss of the boundaries. *Mcp* (miscadastral pigmentation) might delete the boundary between the *iab-4* and *iab-5* regulatory regions, and *Fab* (front abdominal) (H. Gyurkovics et al., pers. comm.) might delete the boundary between *iab-6* and *iab-7*. However, the

3

boundary between the *bxd* and *iab-2* regions seems to be non-essential to the function of either domain. Rearrangement breakpoints that cut near the right end of the *bxd* regulatory region (at about +20 in Fig. 1) leave most of the *bxd* regulation of parasegment 6 undisturbed. Likewise, breaks in the left end of the *iab-2* domain (at about +25 in Fig. 1) leave most of the regulation of parasegment 7 undisturbed. Indeed, flies hemizygous for breaks in either the right end of the *bxd* region or the left end of the *iab-2* region are nearly wild type.

We have also studied the effects of *bxd* and *iab-2* mutations on the patterns of *Ubx* and *abd-A* proteins in embryos. As *bxd* breaks move further to the left, cutting away more and more of the regulatory region from the *Ubx* promoter, more and more of the parasegment 6 pattern of *Ubx* is deleted, but the remaining *bxd* DNA still drives part of the pattern with a proper turn-on in parasegment 6 (our unpublished results). In the same way, *iab-2* breaks remove parts of the pattern, but parts of the pattern remain with proper segmental restriction (F. Karch et al., in prep.). Thus, we cannot point to any one site in either the *bxd* or *iab-2* regulatory regions that is responsible for the segmental restriction.

Transformation Experiments

The P-element transformation system makes possible a more directed analysis of these regulatory regions, although one should ideally use P elements containing an entire segmental domain. We have developed a method for building very large plasmids in *Escherichia coli*, and we have used the method to build P elements containing the entire *bx* or *iab-2* domains (O'Connor et al. 1989). Such large P elements do not transpose at high frequency, and we have not yet gotten such elements into *Drosophila* chromosomes. We have tested P elements containing several smaller fragments of the *bx, bxd, iab-2,* and *iab-3* regulatory regions. Some fragments drive a *Ubx-lacZ* fusion gene in specific cell types but without proper segmental restriction. Other fragments turn on the reporter gene with the proper segmental restriction in early embryos, although the restriction is violated in late embryos. We do not yet know how many different elements exist in any one domain that can confer such segmental restriction, nor what sequences are required to maintain these restrictions in later development. If we can transform flies with the larger P elements, these questions will be easier to address.

4

ACKNOWLEDGMENT

We are grateful to Ed Lewis for providing most of the mutations in the complex and the interpretations of their segmental transformations.

REFERENCES

Celniker, S.E., D.J. Keelan, and E.B. Lewis. 1989. The molecular genetics of the Bithorax Complex of *Drosophila*: Characterization of the products of the Abdominal-B domain. *Genes Dev.* **3:** (in press).

Kaufman, T.C., R. Lewis, and B. Wakimoto. 1980. Cytogenetic analysis of chromosome 3 in *Drosophila melanogaster*: The homeotic gene complex in polytene interval 84A-B. *Genetics* **94:** 115.

Kornfeld, K., R.B. Saint, P.A. Beachy, P.J. Harte, D.A. Peattie, and D.S. Hogness. 1989. Structure and expression of a family of *Ultrabithorax* mRNAs generated by alternative splicing and polyadenylation in *Drosophila*. *Genes Dev.* **3:** 243.

Kuziora, M.A. and W. McGinnis. 1988. Different transcripts of the *Drosophila* Abd-B gene correlate with distinct genetic subfunctions. *EMBO J.* **7:** 3233.

Lewis, E.B. 1978. A gene complex controlling segmentation in *Drosophila*. *Nature* **276:** 565.

O'Connor, M.B., M. Peifer, and W. Bender. 1989. Construction of large DNA segments in *Escherichia coli*. *Science* **244:** 1307.

O'Connor, M.B., R. Binari, L.A. Perkins, and W. Bender. 1988. Alternate RNA products from the Ultrabithorax domain of the bithorax complex. *EMBO J.* **7:** 435.

Peifer, M., F. Karch, and W. Bender. 1987. The bithorax complex-controlling segmental identity. *Genes. Dev.* **1:** 891.

Genes That Control Pattern Formation during Development

M.P. Scott, S. Hayashi, G.M. Winslow, J.E. Hooper, and S. Sonoda

Department of Molecular, Cellular, and Developmental Biology
University of Colorado, Boulder, Colorado 80309

The development of the fruit fly *Drosophila* is controlled by a network of regulatory genes. These genes have been identified by mutations that change the body plan of the embryo or adult. Mutations in segmentation genes alter the number or pattern of body segments in the embryo by eliminating the development of alternate body segments, a part of each segment, or whole groups of segments. Mutations in homeotic genes lead to the development of one part of the body in the pattern of another. For example, legs develop where antennae should be in certain *Antennapedia* (*Antp*) homeotic mutants.

From research in many laboratories, it is now clear that a key to the way in which all these pattern-controlling genes work is by spatially regulated transcription. The earliest acting genes are expressed in broad regions that approximately correspond to the regions of the embryo that are altered when the gene is mutated. Later-acting genes are expressed in increasingly finer patterns and control the development of increasingly finer structures of the embryo. Later-acting genes are controlled by earlier-acting segmentation genes in order to attain the necessary patterns of expression. Homeotic genes are expressed differentially along the anterior-posterior and dorsal-ventral axes of the embryo, and it is this differential expression that directs the parts of the embryo to follow different developmental pathways. Two central problems therefore are to discover how control of temporal and spatial gene expression is accomplished and to learn how the products of the regulatory genes control the formation of morphology.

Control of Transcription by Homeo Domains
Many of the regulatory genes appear to encode transcription factors on the basis that they contain a homeo box, a 183-bp sequence that encodes an evolutionarily conserved 61-amino-acid

7

protein sequence (the homeo domain), which is found in known mammalian transcription factors and is similar to the helix-turn-helix structure that allows sequence-specific DNA binding by bacterial proteins. It seems likely that some of the segmentation genes directly control the transcription of others through their homeo-domain protein products. Using in vitro DNA-binding studies, a cultured cell transfection system, and germ-line transformation of flies, we have investigated the transcription-regulating capabilities of some of the segmentation and homeotic gene protein products, and we have characterized cis-acting DNA sequences on which the proteins act to control transcription. The *Antp* protein has been found to activate transcription in cultured cells if a sequence consisting of a repeating TAA motif, and variations on it, is in *cis* to the target promoter. Such a sequence works if positioned either upstream of or downstream from a promoter and therefore has the properties of an enhancer. In cultured cells, *Antp* protein is capable of activating its own P1 promoter or the promoter of the *Ultrabithorax* (*Ubx*) gene but not several other promoters. Altered forms of the *Antp* protein also activate transcription, including one that has the *Ubx* homeo domain and carboxy-terminal amino acids in place of the corresponding *Antp* sequence.

cis-Acting Control Elements of the *fushi tarazu* Gene

We are analyzing the *cis*-acting control elements of the segmentation gene *fushi tarazu* (*ftz*) in collaboration with C. Wu at the National Institutes of Health. A protein factor, ftzF1, identified in Wu's laboratory, binds to several sites in the *ftz* promoter's upstream region and to sites within the protein-coding sequence of *ftz*. We have tested the effects of altering the strongest of the ftzF1-binding sites on the expression of the *ftz* promoter in vivo using *lacZ* gene fusions. The *ftz* gene is normally expressed in a pattern of seven transverse stripes, whose correct formation is necessary for the development of the normal pattern of body segments. The pattern and intensity of stripes are altered in a *ftz-lacZ* fusion that carries a mutated ftzF1 site. The ftzF1 protein appears to act as an activator of *ftz* in all stripes, but it is especially required for the formation of three of the stripes. In this way, we have begun to identify elements involved in tissue-specific and position-specific control.

Cloning of the *patched* Segmentation Gene

A separate project has been the molecular cloning of the gene *patched* (*ptc*), a segmentation gene that controls pattern formation in part of each body segment. The *ptc* gene is expressed in a complex and dynamic pattern of transverse stripes. Sequence analysis of the gene has revealed that it encodes a product that appears to be a transmembrane protein. Therefore, this gene may function in cell-cell communication rather than in transcriptional regulation.

Spatial Regulation of Homeo Box Gene Expression in *Drosophila*

M. Levine, D. Stanojevic, K. Han, J.L. Manley, and R. Warrior

Department of Biological Sciences, Fairchild Center
Columbia University, New York, New York 10027

Past genetic screens have identified nearly 50 zygotically active genes that control cell fate during early *Drosophila* development (for example, see Nusslein-Volhard and Wieschaus 1980). The vast majority of these genes (almost 40) control the differentiation of the anterior–posterior embryonic body axis, whereas the remaining genes specify dorsal–ventral fates. By now, nearly two-thirds of these regulatory genes have been molecularly cloned and characterized (for review, see Ingham 1988). Surprisingly, over half of them are evolutionarily related by virtue of a 180-bp protein-coding sequence called the homeo box (McGinnis et al. 1984; Scott and Weiner 1984). The homeo box protein domain has been shown to mediate sequence-specific DNA-binding activities, and there is mounting evidence that homeo box proteins control development by modulating gene expression at the level of transcription (for review, see Levine and Hoey 1988).

Nearly every one of the approximately 20 early-acting homeo box genes shows a unique pattern of expression during embryogenesis. In fact, most of the 6000 cells that comprise the early embryo possess a unique combination of active and inactive homeo box genes, and it is thought that these different permutations of homeo box products play a key role in establishing diverse pathways of morphogenesis. A central issue regarding the homeo box gene control of development is a problem of regulation: How do each of these genes come to be expressed in the correct subsets of cells of the early embryo? The importance of this question is underscored by misexpression studies, whereby homeo box genes are ectopically expressed in the "wrong" cells by using heterologous promoters, such as the heat-inducible heat-shock protein hsp70 promoter (Struhl 1985; Schneuwly et al. 1987; Kuziora and McGinnis 1988). Ectopic expression of a given homeo box gene can strongly disrupt the normal spatial organization of the embryo.

A considerable amount of information has been obtained concerning the *trans* control of homeo box gene expression during early development, based on studies that localized various homeo box gene products in different regulatory mutants and mutant combinations (for example, see Ingham 1988). A principal conclusion of these studies is that selective patterns of homeo box gene expression involve a network of cross-regulation, whereby one homeo box gene can influence the expression of others. There is evidence that at least some of these interactions occur at the level of transcription. More specifically, we will discuss a multiswitch mechanism of cross-regulation, whereby divergent homeo box proteins compete for multiple copies of a common DNA-binding site contained within a given target promoter (Han et al. 1989).

Homeo Box Proteins Are Sequence-specific Transcriptional Regulators

DNA-binding studies done with six different full-length homeo box proteins made in bacteria have resulted in the identification of a common 10-bp consensus sequence: TCAATTAAAT (Desplan et al. 1988; Hoey and Levine 1988). These proteins represent a broad spectrum of homeo box types in *Drosophila* with the most divergent of the proteins containing homeo boxes that share less than 50% amino acid identity. In fact, recent studies have shown that the highly divergent octamer-binding protein in mammals, OCT-2, can also recognize the 10-bp consensus sequence in *Drosophila* (Ko et al. 1988). The so-called octamer motif and *Drosophila* consensus sequence are vaguely related, sharing five of eight residues. Thus, it would appear that homeo box proteins bind to a related class of DNA recognition sequences, suggesting the coevolution of homeo box genes and their recognition sequences.

We have used a transient expression assay in *Drosophila* tissue culture cells to demonstrate that the 10-bp DNA-binding motif can mediate the regulatory activities of homeo box proteins in vivo (Han et al. 1989). This assay involves the use of heterologous reporter plasmids containing multiple copies of the 10-bp recognition sequence (between 5–25 copies) attached to the basal promoter of the *Drosophila* metallothionein gene. The expression of these heterologous promoters was monitored using a standard chloramphenicol acetyltransferase assay, when introduced in tissue culture cells along with various homeo-box-expression plasmids (and combinations of these plasmids) driven by the actin 5C promoter. Four of the six proteins tested in this assay, including zerknült (zen), zen-related (z2), fushi tarazu (ftz), and paired (prd), were found to activate specifically the transcription of promoters containing copies of the 10-bp recognition sequence. In the case of zen and z2, activation was particularly striking with 50-fold and 400-fold increases in expression, respectively. The regions of the zen and z2 proteins that are important for this activation have been identified by assaying various mutant proteins. For z2, activation depends on a region of 64 amino acid residues that reside downstream from the homeo domain. This region is highly acidic (12 of 64 residues correspond to aspartate or glutamic acid) and might function in a manner analogous to the so-called acidic domains in prototypic transcriptional activators of yeast, gal4 and GCN4 (for example, see Ma and Ptashne 1987; Hope et al. 1988).

Homeo Box Proteins Act Synergistically

Our primary rationale for establishing this assay system was to reconstruct the combinatorial aspects of homeo box gene activity seen in early embryos. As indicated above, nearly every embryonic cell contains a unique combination of homeo box gene products, and past genetic studies strongly suggest that these products act in concert with one another to control gene expression and guide cell fate. Expression of multiple homeo box proteins in the transient assay system resulted in very striking synergistic activation of reporter plasmids containing multiple copies of the 10-bp recognition sequence. For example, a combination of ftz and zen proteins produced a 950-fold activation of expression, whereas the predicted additive values of the two proteins acting alone are only 16-fold. The most spectacular activation was obtained with a combination of three proteins, zen, ftz, and prd, which produced a 2500-fold stimulation in expression (Han et al. 1989).

Certain homeo box proteins not only failed to participate in this synergistic activation of gene expression, but actually repressed or "quenched" the ability of the activating proteins to function in a synergistic manner. For example, even-skipped (eve) protein reduced the level of activation obtained with a combination of ftz and prd from 220-fold to only 19-fold. Interestingly, this quenching by eve does not depend on DNA bind-

13

ing. Certain mutant *eve* proteins that are unable to bind DNA in vitro are nonetheless able to quench gene expression in this assay. The region of the *eve* protein that participates in quenching maps just downstream from the homeo domain and includes a repeating series of alanine residues.

The results obtained in the transient expression experiments suggest that different homeo box proteins compete for multiple copies of the 10-bp recognition sequence; some combinations of bound homeo box proteins function more efficiently than others in activating transcription. Several mechanisms can be envisioned for the observed synergistic interactions. Our favored working model is that different homeo box proteins make qualitatively distinct interactions with various rate-limiting components of the transcription complex. Perhaps homeo box protein A interacts with a TATA-binding factor, whereas homeo box protein B interacts with the large subunit of polymerase II. Evidence for the importance of protein–protein interactions in the cooperative activation of gene expression stems from studies done with proteins that are unable to bind DNA. For example, strong synergistic activation is obtained with a combination of wild-type *zen* protein and a mutant *ftz* protein, which contains an internal in-frame deletion within its homeo box. This observation indicates that *ftz* need not bind DNA to interact with *zen* cooperatively.

The regulatory activities of homeo box proteins are partly responsible for the restricted expression of certain homeo box genes within tightly localized spatial limits. However, the initiation of the earliest-acting homeo box genes depends on regulatory proteins that do not contain homeo boxes. In particular, the so-called gap genes play a key role in establishing striped expression of different homeotic genes, as well as several members of the pair-rule class of segmentation genes (for example, see Frasch and Levine 1987; Harding and Levine 1988). There are a total of five gap genes, and each of the three that has been cloned and characterized contains the zinc finger DNA-binding motif (for example, see Tautz et al. 1987).

Gap Genes Initiate Stripes
One of the central issues of the segmentation field is how relatively few gap genes possess sufficient informational content to direct the expression of an organized set of pair-rule (and homeotic) stripes. Genetic studies suggest that the establishment of the early seven-stripe pattern of the pair-rule gene *eve*

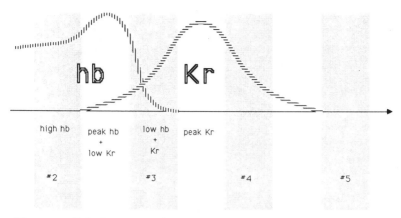

high hb	peak hb	low hb	peak Kr
	+	+	
	low Kr	Kr	

#2 #3 #4 #5

Figure 1 Relative patterns of *hb*, *Kr*, and *eve* expression in an early wild-type embryo. The *hb* and *Kr* proteins are distributed in broad, overlapping gradients, such that six–eight cells express significant levels of both proteins. It is possible that the establishment of an organized set of *eve* stripes (i.e., 2, 3, and 4) involves the threshold response of the *eve* promoter to different levels and combinations of gap proteins.

involves a relatively direct interaction of gap proteins with the *eve* promoter. We have identified a minimal *eve* promoter fragment (<1.7 kb long) that directs the expression of stripe 2 (Harding et al. 1989). This promoter fragment is differentially regulated by the two best characterized gap proteins, *hunchback* (*hb*) and *Krüppel* (*Kr*). *hb* exerts a positive effect on the expression of stripe 2, whereas *Kr* exerts a negative effect.

Immunolocalization studies indicate that both proteins are expressed in broad gradients. Moreover, the domains of *hb* and *Kr* extensively overlap, with at least six cells expressing significant levels of both proteins (see Fig. 1). Full-length *hb* and *Kr* proteins made in bacteria bind to adjacent but distinct sites within the minimal *eve* promoter. Despite the fact that both proteins contain the zinc finger motif, they bind totally distinct classes of DNA-recognition sequences. The *Kr*-binding site includes the core sequence GGGTTAA, whereas the core *hb* sites contain five or six A residues. We discuss these observations in the context of a genetic switch model for the gap-mediated specification of pair-rule stripes.

REFERENCES
Desplan, C., J. Theis, and P.H. O'Farrell. 1988. The sequence-specificity of homeodomain-DNA interaction. *Cell* **54:** 1081.

Frasch, M. and M. Levine. 1987. Complementary patterns of *even-skipped* and *fushi tarazu* expression involve their differential regulation by a common set of segmentation genes in *Drosophila*. *Genes Dev.* 1: 981.

Han, K., M.S. Levine, and J.L. Manley. 1989. Synergistic activation and repression of transcription by *Drosophila* homeo box proteins. *Cell* 56: 573.

Harding, K. and M. Levine. 1988. Gap genes define the limits of *Antennapedia* and *Bithorax* gene expression during early development in *Drosophila*. *EMBO J.* 7: 205.

Harding, K., T. Hoey, R. Warrior, and M. Levine. 1989 Autoregulatory and gap gene response elements of the *even-skipped* promoter of *Drosophila*. *EMBO J.* 8: 1205.

Hoey, T. and M. Levine. 1988. Divergent homeo box proteins recognize similar DNA sequences in *Drosophila*. *Nature* 332: 858.

Hope, I.A., S. Mahadevan, and K. Struhl. 1988. Structural and functional characterization of the short acidic transcriptional activation region of yeast GCN4 protein. *Nature* 333: 635.

Ingham, P.W. 1988. The molecular genetics of embryonic pattern formation in *Drosophila*. *Nature* 335: 25.

Ko, H.-S., P. Fast, W. McBride, and L.M. Staudt. 1988. A human protein specific for the immunoglobulin octamer DNA motif contains a functional homeo box domain. *Cell* 55: 135.

Kuziora, M.A. and W. McGinnis. 1988. Autoregulation of a homeotic selector gene. *Gene* 55: 477.

Levine, M. and T. Hoey. 1988. Homeobox proteins as sequence-specific transcription factor. *Cell* 55: 537.

Ma, J. and M. Ptashne. 1987. Deletion analysis of GAL4 defines two transcriptional activating segments. *Cell* 48: 847.

McGinnis, W., M.S. Levine, E. Hafen, A. Kuroiwa, and W.H. Gehring. 1984. A conserved DNA sequence in homeotic genes of the *Drosophila* Antennapedia and Bithorax complexes. *Nature* 308: 428.

Nusslein-Volhard, C. and E. Wieschaus. 1980. Mutations affecting segment number and polarity in *Drosophila*. *Nature* 287: 795.

Schneuwly, S., R. Klemenz, and W.H. Gehring. 1987. Redesigning the body plan of *Drosophila* by ectopic expression of the homeotic gene *Antennapedia*. *Nature* 325: 816.

Scott, M.P. and A.J. Weiner. 1984. Structural relationships among genes that control development: Sequence homology between the *Antennapedia, Ultrabithroax,* and *fushi tarazu* loci of *Drosophila*. *Proc. Natl. Acad. Sci.* 81: 4115.

Struhl, G. 1985. Near reciprocal genotypes caused by inactivation or indiscriminate expression of the *Drosophila* segmentation gene *ftz*. *Nature* 318: 677.

Tautz, D., R. Lehmann, H. Schnurch, R. Schuh, E. Seifert, A. Kienlin, K. Jones, and H. Jackle. 1987. Finger protein of novel structure encoded by *hunchback*, a second member of the gap class of *Drosophila* segmentation genes. *Nature* 327: 383.

A Homeo Domain Switch Changes the Regulatory Function of the *Deformed* Protein in *Drosophila* Embryos

M.A. Kuziora and W. McGinnis

Department of Molecular Biophysics and Biochemistry
Yale University, New Haven, Connecticut 06511

The unique identity of each segment of *Drosophila melanogaster* is determined early in development by the homeotic selector genes (Lewis 1978; Kaufman et al. 1980). The homeotic selector genes belong to a family of genes involved in *Drosophila* embryonic patterning that are related by a region of homology known as the homeo box (McGinnis et al. 1984a,b; Scott and Weiner 1984). All of the homeotic selectors are members of the *Antennapedia* (*Antp*) class of the homeo box gene family, a subclass whose members share 60–90% identity in the amino acid sequences of their respective homeo domains (Regulski et al. 1985). The carboxyl half of homeo domains contains structural similarity to the helix-turn-helix motif common to many prokaryotic DNA-binding proteins (Laughon and Scott 1984; Otting et al. 1988). The presence of a DNA-binding domain in homeotic proteins is consistent with their proposed function as transcriptional regulators of other homeotic selector genes and of downstream genes required for elaboration of segmental morphology.

Homeo-domain-containing proteins do bind DNA in a sequence-specific manner in vitro; indeed, in vitro assays indicate many homeo domain proteins bind in vitro to the same or very similar sequences (Hoey and Levine 1988; Desplan et al. 1988).

Transient expression assays carried out in tissue culture cells have also demonstrated that although homeo-domain-containing proteins can function as transcription regulators in vivo (presumably by directly binding to promoter elements that were fused upstream of reporter genes), they do so in a relatively nonspecific manner (Jaynes and O'Farrell 1988; Thali et

17

al. 1988). The promiscuous DNA-binding activity of homeo-domain-containing proteins and their generic regulatory abilities in tissue culture cells are difficult to reconcile with their proposed role in regulating specific genes required for unique segmental morphology.

We have demonstrated previously that ectopic expression of the homeotic selector genes *Deformed* (*Dfd*), using a heat-shock promoter fused to *Dfd*-coding sequences (*hsDfd*), transforms head and thoracic segments toward a maxillary identity (Kuziora and McGinnis 1988). The observed transformations were dependent on the activation and persistent expression of the endogenous *Dfd* gene through an autoregulatory circuit. This experimental system provides us with a functional assay in which experimentally altered *Dfd* proteins can be tested for their ability to activate autoregulation of the endogenous *Dfd* gene.

To begin to test where the regulatory specificity of homeo proteins maps within their primary sequence, we have replaced the homeo domain of *Dfd* with the homeo domain of *Ultrabithorax* (*Ubx*), a homeotic selector that specifies the third thoracic–first abdominal segment identity. Our results show that the chimeric protein has lost the ability to activate *Dfd* autoregulation. Instead, the chimeric protein now activates *Antp* gene expression in head segment primordia. As a result, head segments are transformed toward a thoracic identity.

RESULTS

To test the requirement for specific homeo domain sequences for determining homeotic regulatory functions, we precisely replaced the homeo box of *Dfd* in a *hsDfd* gene with the homeo box from *Ubx*. We chose *Ubx* for the substitution for two reasons: (1) It is a homeotic selector gene not involved in head development; and (2) when *Ubx* is ectopically expressed in transformed flies carrying a *Ubx* gene under control of the heat-shock promoter, head and thoracic segments are transformed to the first abdominal segment (G. Struhl, pers. comm.). The homeo box switch involved an exact replacement of 66 amino acid residues of *Dfd* with 66 amino acid residues of *Ubx* with no insertions or deletions at the boundaries of the substituted region. It is also of interest to note that the amino acid sequence of the "recognition" helix, a region of the helix-turn-helix motif demonstrated previously to be important in determining DNA–protein contacts (Wharton and Ptashne 1985; Anderson et al. 1987), is identical in the two homeo domains.

A P-element transposon carrying the *hsDfd/Ubx* chimeric gene was used to transform a *rosy* strain. We first tested the capability of the *hsDfd/Ubx* switch protein to activate the *Dfd* autoregulatory circuit. Embryos from a *hsDfd/Ubx* switch strain were heat shocked during the cellular blastoderm stage and allowed to recover for 1 or 5 hours at 25°C. In embryos containing the "normal" *hsDfd* gene, the activation of the endogenous *Dfd* gene can be immunologically detected in stripes of expression that are restricted to the ventral posterior ectoderm of each segment after 5 hours recovery from heat shock (Kuziora and McGinnis 1988). The ectopic expression of the *hsDfd/Ubx* chimeric protein cannot initiate this ectopic activation of a *Dfd* autoregulatory circuit.

Ectopic expression of homeotic selector genes during development often alters segmental identity as reflected in cuticular morphology. For example, ectopic expression of *Dfd* during the cellular blastoderm stage causes transformations of head and thoracic segments toward a maxillary identity (Kuziora and McGinnis 1988). The observed transformations were dependent on the activation of the autoregulated expression of the endogenous *Dfd* gene in other body segments. We examined cleared cuticles of embryos ectopically expressing the *hsDfd/Ubx* switch gene to test for alternative phenotypes.

Heat-shock-induced misexpression of the *Dfd/Ubx*-switch-protein treatment is lethal and results in embryos with strong transformations of head segments to thoracic identities, as evidenced by the development of thoracic denticle belts in place of the normal head structures. The phenotypic transformations observed as the result of ectopic expression of the *hsDfd/Ubx* switch protein bore remarkable similarity to cuticular transformations resulting from misexpression of the *Antp* gene (Gibson and Gehring 1988). We therefore tested for ectopic expression of the *Antp* gene by heat shocking embryos from the *hsDfd/Ubx* switch strain at the cellular blastoderm, allowing them to recover for 1 or 3 hours, and then fixing and sectioning the embryos. The sectioned embryos were then hybridized in situ with an *Antp* antisense RNA probe. Embryos misexpressing the *hsDfd/Ubx* chimeric gene activate *Antp* transcription in the head segments, a region far outside its normal domain of expression in the thorax. No ectopic activation of *Antp* was ob-

19

served in nonswitched *hsDfd* embryos given similar heat-shock treatment.

DISCUSSION

Our results suggest that the differences in the amino acid sequence of even closely related homeo domains contain much of the genetic regulatory specificity of the entire homeo protein. Global expression of the normal *Dfd* protein induces ectopic activation of the endogenous *Dfd* gene and subsequent transformation of thoracic segments toward a head segment identity. In contrast, global expression of the *Dfd/Ubx* chimeric protein leads to ectopic activation of the *Antp* gene and a subsequent transformation of head segments toward a thoracic identity. The change in genetic regulatory specificity, and the resulting effects on embryonic morphology, of the normal and switched homeo proteins is shown in Figure 1.

Replacement of the *Dfd* homeo domain with the *Ubx* homeo domain has resulted in a chimeric protein with a new regulatory property, the ability to activate *Antp* expression. Why is

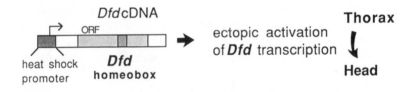

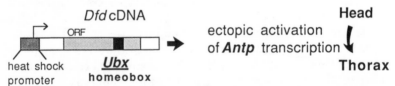

Figure 1 Embryonic regulatory functions of the *Dfd* and the *Dfd/Ubx* switch proteins. The heat-shock-induced expression of *Dfd* protein in all embryonic cells results in the ectopic activation of the normal chromosomal *Dfd* transcription unit. This in turn results in the partial transformation of thoracic segments into head (maxillary) segments. The substitution of the *Ubx* homeo box for the *Dfd* version results in the loss of autoregulation and the gain of a positive regulatory effect on the *Antp* transcription unit. This in turn results in a partial transformation of head segments into thoracic segments.

Antp activated as a result of misexpression of the chimeric protein? One possible explanation is that the normal genetic regulatory specificity of *Ubx* has been transferred to the *Dfd* protein with the *Ubx* homeo box, but the regulatory effect on the *Antp* promoter has been changed from negative to positive. *Ubx* normally acts as a repressor of *Antp* expression in PS6 (Hafen et al. 1984), and a *Ubx* protein isoform has been shown to bind to the *Antp* promoter region in vitro (Beachy et al. 1988). Most eukaryotic regulatory proteins are composed of at least two domains: one that recognizes and binds to specific DNA sequences in the promoter of the target gene and a second domain that acts as a general transcription activator (Hope et al. 1988). Activation domains often are rich in acidic amino acid residues. A scan of the sequences of *Ubx* protein isoforms reveals that none contain a domain consisting of predominantly acidic amino acid residues. The *Dfd* protein, however, contains a region of acidic residues between the amino end of the protein and the homeo domain (Regulski et al. 1987). Thus, the specificity of the *Ubx* homeo domain for the *Antp* promoter, combined with the activating effect of the *Dfd* acidic region, may account for the altered regulatory function of the *Dfd/Ubx* switch protein.

ACKNOWLEDGMENTS

We would like to thank the National Science Foundation and the National Institutes of Health for their generous support of our research.

REFERENCES

Anderson, J.E., M. Ptashne, and S.C. Harrison. 1987. Structure of the repressor-operator complex of bacteriophage 434. *Nature* **326:** 846.

Beachy, P.A., M.A. Krasnow, E.R. Gavis, and D.S. Hogness. 1988. An *Ultrabithorax* protein binds sequences near its own and the *Antennapedia* P1 promoters. *Cell* **55:** 1069.

Desplan, C., J. Theis, and P.H. O'Farrell. 1985. The *Drosophila* developmental gene, *engrailed*, encodes a sequence-specific DNA binding activity. *Nature* **318:** 630.

———. 1988. The sequence specificity of homeodomain-DNA interaction. *Cell* **54:** 1081.

Gibson, G. and W.J. Gehring. 1988. Head and thoracic transformation caused by ectopic expression of *Antennapedia* during *Drosophila* development. *Development* **102:** 657.

Hafen, E., M. Levine, and W.J. Gehring. 1984. Regulation of *Antennapedia* transcript distribution by the bithorax complex in *Drosophila. Nature* **307:** 287.

Hoey, T. and M. Levine. 1988. Divergent homeo box proteins recognize similar DNA sequences in *Drosophila*. *Nature* **332:** 858.

Hope, I.A., S. Mahadevan, and K. Struhl. 1988. Structural and functional characterization of the short acidic transcriptional activation region of yeast GCN4 protein. *Nature* **333:** 635.

Jaynes, J.B. and P.H. O'Farrell. 1988. Activation and repression of transcription by homeodomain-containing proteins that bind to a common site. *Nature* **336:** 744.

Kaufman, T.C., R. Lewis, and B. Wakimoto. 1980. Cytogenetic analysis of chromosome 3 in *Drosophila melanogaster*: The homeotic gene complex in polytene chromosome interval 84A-B. *Genetics* **94:** 115.

Kuziora, M.A. and W. McGinnis. 1988. Autoregulation of a *Drosophila* homeotic selector gene. *Cell* **55:** 477.

Laughon, A. and M.P. Scott. 1984. Sequence of a *Drosophila* segmentation gene: Protein structure homology with DNA-binding proteins. *Nature* **310:** 25.

Lewis, E.B. 1978. A gene complex controlling segmentation in *Drosophila*. *Nature* **276:** 565.

McGinnis, W., R.L. Garber, J. Wirz, A. Kuroiwa, and W.J. Gehring. 1984a. A homologous protein-coding sequence in *Drosophila* homeotic genes and its conservation in other metazoans. *Cell* **37:** 403.

McGinnis, W., M. Levine, E. Hafen, A. Kuroiwa, and W.J. Gehring. 1984b. A conserved DNA sequence found in homeotic genes of the *Drosophila* Antennapedia and Bithorax complexes. *Nature* **308:** 428.

Otting, G., Y. Qian, M. Muller, M. Affolter, W. Gehring, and K. Wuthrich. 1988. Secondary structure determination for the *Antennapedia* homeodomain by nuclear magnetic resonance and evidence for a helix-turn-helix motif. *EMBO J.* **7:** 4305.

Regulski, M., N. McGinnis, R. Chadwick, and W. McGinnis. 1987. Developmental and molecular analysis of *Deformed*: A homeotic gene controlling *Drosophila* head development. *EMBO J.* **6:** 767.

Regulski, M., K. Harding, R. Kostriken, F. Karch, M. Levine, and W. McGinnis. 1985. Homeo box genes of the Antennapedia and Bithorax complexes of *Drosophila*. *Cell* **43:** 71.

Scott, M.P. and A. Weiner. 1984. Structural relationships among genes that control development: Sequence homology between the *Antennapedia, Ultrabithorax,* and *fushi tarazu* loci of *Drosophila*. *Proc. Natl. Acad. Sci.* **81:** 4115.

Thali, M., M.M. Muller, M. DeLorenzi, P. Matthias, and M. Bienz. 1988. *Drosophila* homeotic genes encode transcriptional activators similar to mammalian OTF-2. *Nature* **336:** 598.

Wharton, R.P. and M. Ptashne. 1985. Changing the binding specificity of a repressor by redesigning an α-helix. *Nature* **316:** 601.

Transcriptional Regulation by Homeotic Gene Products in Cultured *Drosophila* Cells and In Vitro

E. Parker, B. Johnson, E.E. Saffman, and M.A. Krasnow

Department of Biochemistry, Stanford University
Stanford, California 94305

The protein products of the homeotic genes select and maintain segment identity during *Drosophila melanogaster* development, and each gene is associated with the identity of segments or parasegments in a particular anatomical region. These genes constitute a key intermediate stage of a genetic regulatory hierarchy in which early-acting genes involved in segmentation of the animal, such as *fushi tarazu (ftz)*, ensure the initial region-specific activation of *Ultrabithorax (Ubx)*, *Antennapedia (Antp)*, and other homeotics (for review, see Akam 1987). Later in development, the patterns of expression of the homeotics are refined by cross-regulation and autoregulation. For example, *Ubx* gene products are required to repress *Antp* expression in the posterior thoracic-anterior abdominal region of the embryo, and they are similarly required for high-level *Ubx* expression in the visceral mesoderm (Hafen et al. 1983; Bienz and Tremml 1988). It is also thought that *Ubx* and other homeotics regulate an as yet undetermined set of genes ("realizator" or "downstream" genes) that elaborate the form and function of each segment. We are studying the activities of *Ubx* proteins (UBX) in a *D. melanogaster* cell culture system and in vitro to elucidate the biochemical mechanisms by which they regulate their targets.

A purified UBX protein produced in *Escherichia coli* has sequence-specific binding activity, and it interacts with clusters of binding sites located within several hundred base pairs downstream from the *Ubx* and the *Antp* P1 transcription start sites and approximately 6 kb upstream of *Antp* P1 (Beachy et al. 1988). The binding-site clusters range in size from approximately 40 to 90 bp, and they contain tandem repeats of the trinucleotide TAA. A cloned synthetic oligonucleotide, $(TAA)_4$, is sufficient for UBX binding, and thus the naturally occurring clusters likely accommodate several UBX proteins. In addition, each binding-site cluster is accompanied by another cluster no farther than 140 bp away, so that the binding sites near the *Ubx* (P_{Ubx}) and *Antp* P1 ($P_{Antp\ P1}$) promoters appear in pairs of clusters.

Transcriptional Regulation by a UBX Protein in Cultured Cells and In Vitro

The discovery of UBX binding sites near the promoters of target genes suggested that UBX proteins might directly regulate target gene transcription. We have demonstrated that UBX proteins are DNA-binding transcriptional regulators in vivo and in vitro. In a *D. melanogaster* cell culture cotransfection assay, UBX proteins activated transcription of a heterologous promoter (the alcohol dehydrogenase distal promoter, P_{Adh}) more than 100-fold when two or more copies of a consensus binding-site cluster, $(TAA)_{12}$, were inserted near the transcription start site (Krasnow et al. 1989). Activation was not dependent on the orientation of the inserted sequences, and they could function when placed several hundred base pairs upstream of the transcription start site. UBX proteins are therefore transcriptional regulators in cultured cells, and they can act as enhancer-activating proteins. To determine whether this transcriptional regulation is direct, the interaction was reconstructed in vitro using nuclear extracts of cultured cells and purified UBX protein produced in *E. coli*. Purified UBX protein stimulated transcription of P_{Adh} templates severalfold, and the stimulation was strictly dependent on the presence of UBX binding sites.

The effect of UBX proteins on natural *D. melanogaster* promoters has also been examined in the cell culture cotransfection system (Krasnow et al. 1989). Although UBX proteins had only small effects on most promoters tested, they stimulated transcription of P_{Ubx} from 10- to 30-fold and reduced expression from $P_{Antp\ P1}$ to a similar extent. Thus, UBX proteins are both transcriptional activators and repressors, and some of these activities in cultured cells appear to mimic UBX activities in the developing embryo.

Roles of UBX Protein Binding Sites in Regulation

How does a UBX protein activate some targets and repress others? Part of the answer may be that not all UBX binding sites are functionally equivalent. At P_{Ubx}, there are two clusters of binding sites downstream from the transcriptional start site, between +41 and +88 and between +218 and +306 (Beachy et al. 1988). The distal downstream cluster is required for activation since deleting this region or replacing it with a random sequence dramatically reduced activation (Krasnow et al. 1989).

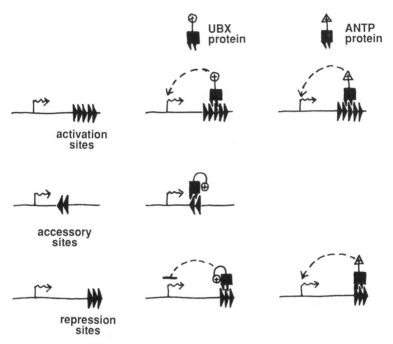

Figure 1 Possible roles of UBX protein binding sites. Each arrowhead indicates a UBX protein binding site positioned near a promoter region. The binding sites are shown clustered and positioned downstream from the transcription start site because these are common features of naturally occurring sites at P_{Ubx} and $P_{Antp\ P1}$. The unoccupied sites are shown at the left along with their designations with respect to their UBX regulatory function. The center and righthand diagrams show the sites occupied by a UBX protein (center) or an ANTP protein (right) and the consequent regulatory effect (dashed line). The two proteins are drawn as DNA-binding domains (closed squares) linked to transcriptional activating domains (circle or triangle surrounding a "+"), since there is some evidence for such a bipartite structure (Krasnow et al. 1989; Winslow et al. 1989). In the models, different binding-site clusters confer different regulatory effects on a given protein (e.g., center column: activation, no effect, and repression by UBX), and a given cluster can mediate different effects depending on the bound protein (e.g., bottom row: repression by UBX and activation by ANTP).

146,773

In contrast, replacing the promoter-proximal cluster with a random sequence substantially increased the UBX stimulatory effect. These results imply that there are at least two functional classes of UBX binding sites (Fig. 1): *activation* clusters, which increase the activity of a nearby promoter in response to UBX; and either *accessory* clusters, which have little effect on their own but may influence the response of other elements to UBX proteins, or *repression* clusters, which decrease the activity of a nearby promoter in response to UBX.

There is evidence that there are repression clusters downstream from the *Antp* P1 transcription start site. Deletion of a downstream region containing two UBX-binding-site clusters reduced repression by UBX from 27- to 8-fold. Furthermore, inserting this region upstream of another promoter conferred a modest repression by UBX. Since this downstream region does not appear to positively influence basal level promoter activity, UBX proteins may repress via this region by inactivating or interfering with factors required for basal level expression that function outside of the region.

In addition to indicating the downstream repression element at $P_{Antp\ P1}$, the results described above also demonstrated that there are additional UBX-repression elements at $P_{Antp\ P1}$ because UBX proteins were still able to repress significantly promoter constructs that lack this downstream region. The UBX binding sites located 6 kb upstream of $P_{Antp\ P1}$ are not important because deletion of this region did not reduce repression. Indeed, $P_{Antp\ P1}$ constructs with all known UBX binding sites deleted were still repressed by UBX. Thus, either cryptic UBX binding sites exist in these constructs, or UBX proteins are capable of repressing transcription by a mechanism that does not require direct binding to target sequences. There is precedent for binding-independent regulatory activity among other DNA-binding transcriptional regulators, such as the estrogen receptor and the yeast *GAL4* protein, which can repress certain target genes apparently by interacting with (and inactivating) proteins required for target gene transcription (Adler et al. 1988; Gill and Ptashne 1988).

Transcriptional Regulation by Other Homeo Domain Proteins

The discovery that proteins encoded by other homeotic genes, such as an *Antp* protein (ANTP), can also act as transcriptional activators in cultured cells strongly supports the notion that these proteins, which all contain a similar 60-amino-acid motif called the homeo domain, are all DNA-binding transcriptional regulatory proteins (Thali et al. 1988; Winslow et al. 1989). Since 20 or so genes with very different roles in the developmental regulatory hierarchy also encode homeo domain proteins (for review, see Akam 1987; Scott et al. 1989), and since some of these are now known to function as transcriptional regulators in cultured cells (Jaynes and O'Farrell 1988; Driever and Nusslein-Volhard 1989; Han et al. 1989), it is very likely that the core of the hierarchy is a cascade of transcriptional control proteins.

An interesting complexity in this cascade has arisen from DNA-binding experiments and cell culture studies, which have shown that several different homeo domain proteins can act via the same regulatory sites. For example, UBX, ANTP, and a *ftz* protein can each activate transcription through the distal downstream cluster at P_{Ubx} (Krasnow et al. 1989; Winslow et al. 1989), and all three proteins can activate the P_{Adh} constructs that contain the consensus UBX binding sites. Some promoters, however, respond differently to these proteins. For example, $P_{Antp \ P1}$ is repressed by UBX but stimulated by ANTP in the cotransfection assay. Also, some synthetic target genes are activated by one homeo domain protein and unaffected by a second, but the second protein can quench the effect of the first (Jaynes and O'Farrell 1988; Han et al. 1989). One reasonable explanation of these results is that different homeo domain proteins can interact with some of the same target sequences, but the particular sequence or the organization of binding sites in a binding-site cluster (or a pair of clusters) dictates the regulatory effect of the bound protein (Fig. 1). Different sequences could have different effects on various homeo domain proteins (Fig. 1), and synergistic, antagonistic, and perhaps unexpected regulatory consequences might result in the developing animal if more than one homeo domain protein binds simultaneously or sequentially to a binding-site cluster.

There are several major challenges ahead in unraveling the developmental regulatory hierarchy at the biochemical level. First, the mechanisms by which a single homeo domain protein such as UBX can exert opposite transcriptional effects when bound to different regulatory sites must be elucidated. Second, the mechanisms by which different homeo domain proteins bound to the same region can exert different effects must be determined. Third, the binding specificities of the various homeo domain proteins must be characterized further, particularly with respect to the role this selectivity plays in dictating target gene specificity in the developing animal. Fourth, regulatory mechanisms of homeo domain proteins that do not involve direct DNA binding must be characterized and evaluated. And finally, it must be determined how different homeo domain proteins interact with each other and with other components of the transcriptional machinery to exert their diverse regulatory effects.

ACKNOWLEDGMENTS

B.J. was supported by the Medical Scientist Training Program of the National Institutes of Health (NIH); E.E.S. was supported by a training grant from National Institute of General Medical Sciences; and M.A.K. is a Lucille P. Markey Scholar in Biomedical Science. This work was supported by grants from the Lucille P. Markey Charitable Trust and the NIH (to M.A.K.).

REFERENCES

Adler, S., M.L. Waterman, X. He, and M.G. Rosenfeld. 1988. Steroid-receptor mediated inhibition of rat prolactin gene expression does not require the receptor DNA-binding domain. *Cell* **52:** 685.

Akam, M. 1987. The molecular basis for metameric pattern in the *Drosophila* embryo. *Development* **101:** 1.

Beachy, P.A., M.A. Krasnow, E.R. Gavis, and D.S. Hogness. 1988. An *Ultrabithorax* protein binds sequences near its own and the *Antennapedia* P1 promoters. *Cell* **55:** 1069.

Bienz, M. and G. Tremml. 1988. Domain of *Ultrabithorax* expression in *Drosophila* visceral mesoderm from autoregulation and exclusion. *Nature* **333:** 576.

Driever, W. and C. Nusslein-Volhard. 1989. The *bicoid* protein is a positive regulator of *hunchback* transcription in the early *Drosophila* embryo. *Nature* **337:** 138.

Gill, G. and M. Ptashne. 1988. Negative effect of the transcriptional activator GAL4. *Nature* **334:** 721.

Hafen, E., M. Levine, and W.J. Gehring. 1983. Regulation of *Antennapedia* transcript distribution by the bithorax complex in *Drosophila. Nature* **309:** 287.

Han, K., M.S. Levine, and J.L. Manley. 1989. Synergistic activation and repression of transcription by *Drosophila* homeobox proteins. *Cell* **56:** 573.

Jaynes, J.B. and P.H. O'Farrell. 1988. Activation and repression of transcription by homeodomain-containing proteins that bind a common site. *Nature* **336:** 744.

Krasnow, M.A., E.E. Saffman, K. Kornfeld, and D.S. Hogness. 1989. Transcriptional activation and repression by *Ultrabithorax* proteins in cultured *Drosophila* cells. *Cell* **57:** 1031.

Scott, M.P., J.W. Tamkun, and G.W. Hartzell. 1989. The structure and function of the homeodomain. *Biochim. Biophys. Acta.* (in press).

Thali, M., M.M. Muller, M. DeLorenzi, P. Matthias, and M. Bienz. 1988. *Drosophila* homeotic genes encode transcriptional activators similar to mammalian OTF-2. *Nature* **336:** 598.

Winslow, G.M., S. Hayashi, M. Krasnow, D.S. Hogness, and M.P. Scott. 1989. Transcriptional activation by the *Antennapedia* and *fushi tarazu* proteins in cultured *Drosophila* cells. *Cell* **57:** 1017.

seventess, a Protein Tyrosine Kinase Receptor for Positional Information in the Developing *Drosophila* Eye

D.D.L. Bowtell, M.A. Simon, B.E. Kimmel, and G.M. Rubin

Howard Hughes Medical Institute and Department of Biochemistry
University of California, Berkeley, California 94720

The formation of an ordered tissue requires the differentiation of distinct cell types and their organization into integrated structures. Crucial to this process is the generation, reception, and interpretation of positional information by cells in the developing tissue. The eye of *Drosophila* is well suited to an investigation of the processes of cell-cell interaction that allow the formation of a precise cellular pattern of differentiated cells from an initially unpatterned epithelium. The adult *Drosophila* eye consists of an array of approximately 800 hexagonal units or ommatidia, which are composed of a restricted number of specialized cell types. Each ommatidium contains a central core of eight photoreceptors, which are surrounded by pigment cells and lens-secreting cone cells. Overt differentiation and arrangement of these cells into ommatidial precursors begins during late larval development in an initially unpatterned epithelial monolayer, the eye imaginal disc. Ommatidial assembly begins at the posterior edge of the disc and progresses anteriorly over about 2 days. As a result, eye imaginal discs removed during this period contain a smoothly graded sequence of developing ommatidia, with the most mature ommatidia at the posterior edge (for review, see Tomlinson 1988). Mosaic analysis has previously shown that cells differentiate and are recruited into ommatidia in a lineage-independent manner (Ready et al. 1976; Lawrence and Green 1979), indicating that inductive cell-cell interactions govern the process of ommatidial development. This notion is supported by the phenotype of mutants in which ommatidial development is interrupted at specific points (Harris et al. 1976; Reinke and Zipursky 1988; Tomlinson et al. 1988). One of these mutants is *sevenless* in which absence of the gene results in failure to form R7 photoreceptors (Harris et al. 1976; Campos-Ortega et al. 1979).

Molecular and Protein Analysis of *sevenless*

The *sevenless* gene has been cloned (Banerjee et al. 1987; Hafen et al. 1987), and the phenotype has been rescued completely by germ-line transformation of a 16.3-kb genomic fragment (Hafen et al. 1987; Basler and Hafen 1988; Bowtell et al. 1988). We have determined the structure of the transcribed region by sequencing this fragment and corresponding cDNA. The transcript of the *sevenless* locus is approximately 8.6 kb long and can be detected in the larval eye disc, adult head, and body. Translation of the compiled cDNA sequence (Bowtell et al. 1988) revealed an open reading frame of 7677 nucleotides, which included at the carboxyl terminus a domain with a high degree of homology with several protein tyrosine kinase receptors, particularly *c-ros*. Hydrophobicity analysis revealed two distinct hydrophobic regions, one approximately 500 residues from the carboxyl terminus and a second near the amino terminus (values +2.6 and +2.5, respectively; values greater than +1.6 have a high probability of spanning the membrane). The latter is 56 amino acids from the nearest methionine codon and conforms to the limited consensus for amino-terminal anchor sequences (von Heijne 1986). The presence of two putative transmembrane domains suggests a structure in which there is a cytoplasmic-facing tyrosine kinase domain and an extracellular domain loop of approximately 2000 residues, anchored at the amino and carboxyl termini. The putative extracellular domain lacks the cysteine-rich domains characteristic of some other protein tyrosine kinase receptors. It shares about 32% sequence similarity with *c-ros* over 500 amino acids of the presently available *c-ros* sequence (Bowtell et al. 1988), suggesting that *c-ros* may be a vertebrate homolog of *sevenless*.

To analyze the structure of the protein further, we stably transfected a *Drosophila* tissue culture cell line (Schneider L2) with a construct in which the *sevenless* coding region was fused to the actin 5C promoter (M.A. Simon et al., in prep.). Cell extracts were analyzed using antisera raised to synthetic peptides to the carboxy-terminal tail and to fused proteins from the putative extracellular domain of *sevenless*. Pulse-chase labeling and immunoprecipitation experiments showed that the *seventess* protein is synthesized as a single polypeptide chain of approximately 280 kD and then is cleaved into a 220-kD α-chain and a 60-kD β-chain (M.A. Simon et al., in prep.). The β-chain encompasses the kinase and adjacent transmembrane domain and associates noncovalently with the α-chain because SDS can disrupt the association of the two chains. Immunofluorescence, using antisera to specific regions and intact versus fixed and permeabilized cells, has confirmed the predicted domain topology of the protein. However, we have not yet determined whether the amino terminus of the α-chain is anchored in the membrane. Using membrane preparations, we have shown that the *seventess* protein is capable of both autophosphorylation and of phosphorylating an exogenous substrate (enolase) on tyrosine.

Expression of *seventess* during Eye Development

The pattern of *seventess* protein expression was examined in eye discs by light and electron microscopy using the above antibodies (Tomlinson et al. 1987). A single eye disc removed from late third instar larvae contains a smoothly graded series of developing ommatidia extending about halfway across the disc, from the most mature at the posterior edge to the earliest stages in the middle. The cells that will comprise the eight photoreceptors and the lens-secreting cone cells can be individually identified within each ommatidia as they join the cluster and differentiate (for review, see Tomlinson 1988). Therefore, it was possible to determine the cell-specific and temporal appearance of *seventess* expression during the sequence of ommatidial development (summarized in Fig. 5 of Tomlinson et al. 1987). Expression of *seventess* protein was first detected at the boundary between the unpatterned epithelium and the earliest discernable ommatidial stage. Staining in the developing ommatidia was limited to a transient expression in the presumptive R3, R4, and R7 photoreceptor cells and in the cone cells. The majority of the protein in these cells was localized in microvilli at the apical surface of the epithelium. However, membrane staining was also detected slightly deeper in the cells, at the level of the adherence junction, in the three expressing photoreceptor cells. Here, we saw an accumulation of the stain in cells R3, R4, and R7 at the point where their membranes contacted the central R8 cell but not where they abutted other cells. The cone cells, which do not contact R8, did not show this restricted membrane localization. The restricted accumulation of *seenless* protein on the membrane of R3, R4, and R7 may represent a cell-cell ligand receptor interaction in which cell R8 expresses the *seenless* ligand.

Generation and Function of the *seenless* Pattern of Expression

Identification of the complex pattern of *seenless* expression in the developing eye disc raised two related questions: How was the pattern of expression generated, and what was its role in the positional information that specified the R7 cell? To address the first question, we coupled the *seenless* promoter to the reporter gene *lacZ* and examined the pattern of β-galactosidase expression in the developing eye disc. The genomic fragment that is capable of rescuing the *seenless* phenotype includes approximately 950-bp 5′ of the major transcriptional start site. However, transcriptional fusions with *lacZ* made using this 5′-flanking sequence were not expressed in transgenic larvae, nor were translational fusions which also included the long (~950 nucleotides) untranslated leader. Results obtained with *seenless* minigene constructs suggested that regulatory sequences might lie within introns 2–7, at least 3-kb 3′ to the transcription start sites (Bowtell et al. 1988). The presence of enhancer sequences within this region was confirmed when genomic fragments from this region were placed 5′ of the *seenless* promoter/*lacZ* cassette (Bowtell et al. 1989b). Examination by light and electron microscopy of the pattern of β-galactosidase expression in developing ommatidia expressing these fully active constructs revealed a pattern of expression indistinguishable from the previously defined *seenless* protein pattern. This indicates that transcriptional regulation of the *seenless* gene can completely account for the complex pattern of protein expression seen during eye development. We found that the enhancer sequences within the body of the gene are capable of directing the correct pattern and level of expression on the heterologous promoters heat-shock 70 and actin 5C. We are presently defining the positions of these regulatory elements more accurately as a route to isolating transcription factors that interact with the *seenless* gene during eye development.

To address the issue of the role of the pattern of *seenless* expression in R7 cell development, we replaced the normally restricted pattern of *seenless* expression with one where a partial cDNA was expressed ubiquitously under the control of the heat-shock promoter (Bowtell et al. 1989a). The use of an inducible promoter allowed us to investigate the period of requirement of *seven less* for R7 cell development and whether it was required for the maintenance of this cell. In addition, although it is known from mosaic analysis that *seven less* is not required for the development of other photoreceptor cells, it is not known whether the normally restricted pattern of *seven less* expression is important to preventing other developing photoreceptor cells from receiving inappropriate information. We were able to address this question by forcing *seven less* expression in cells that do not normally express the protein. Transient expression of the protein during eye development in a *seven less* larvae resulted in a stripe of rescued ommatidia in otherwise mutant eyes. This showed that there is only a brief period during ommatidial development during which assembling ommatidia require and can respond to the *seven less* protein to produce the R7 cell. Ubiquitous expression of *seven less* had no deleterious effect on development of the animal in general or on other photoreceptor cells in particular.

CONCLUDING REMARKS
We have determined that structure of the *seven less* gene and have characterized its protein product, showing it to be a noncovalently linked α–β heterodimeric cell-surface receptor with protein tyrosine activity. Compared with other protein tyrosine kinase receptors so far characterized, the *seven less* protein is unusual in several respects. It has an unusually large (~2000 amino acid) extracellular domain with two potential transmembrane domains, which together may result in an extracellular loop structure. The protein is involved in the specification of a single cell type, as opposed to a general requirement for cell growth, and is expressed in a dynamic pattern in the developing eye in which substantial differences in protein expression are apparent in ommatidia only 1–2 hours apart in the developmental sequence. Rapid changes in expression of surface receptors through posttranscriptional receptor down-regulation have been observed previously (for review, see Sibley et al. 1987). It is of interest that the dynamic pattern of expression of this receptor can be obtained through transcriptional regulation.

By expressing *seven less* under the control of the heat-shock promoter in the developing eye disc, we have demonstrated a limited period in which the protein is required for R7 cell development and that the restricted wild-type pattern of expression is not required for the development of either the R7 cell or other retinal cells. Together, these results suggested that, although *seventess* transmits a signal necessary for the specification of the R7 cell, the normally complex pattern of *seventess* expression within the developing ommatidium is not part of the positional information required for R7 specification. As a result, we believe that the specificity of the cell signaling mediated by the *seventess* protein must reside in some combination of the temporal and spatial distributions of the *seventess* ligand or other components of the *seventess* signaling pathway. Finally, our findings caution that, in the absence of additional information, care is required when drawing inferences about the role of a gene's pattern of expression during development.

ACKNOWLEDGMENTS
We would like to thank Tom Lila, Dave Hackett, and Chris Ginocchio for help with transformation of P element constructs. D.D.L.B. is a C.J. Martin fellow and M.A.S. is a Damon Runyon-Walter Winchell fellow (grant DRG-926).

REFERENCES
Banerjee, U., P.J. Renfranz, J.A. Pollock, and S. Benzer. 1987. Molecular characterization and expression of *seventess,* a gene involved in neuronal pattern formation in the *Drosophila* eye. *Cell* **49:** 281.

Basler, K. and E. Hafen. 1988. Control of photoreceptor cell fate by the *seventess* protein requires a functional tyrosine kinase domain. *Cell* **54:** 299.

Bowtell, D.D.L., M.A. Simon, and G.M. Rubin. 1988. Nucleotide sequence and structure of the *seventess* gene of *Drosophila melanogaster. Genes Dev.* **2:** 620.

———. 1989a. Ommatidia in the developing *Drosophila* eye require and can respond to *seventess* for only a restricted period. *Cell* **56:** 931.

Bowtell, D.D.L., B.E. Kimmel, M.A. Simon, and G.M. Rubin. 1989b. Regulation of the complex pattern of *seventess* expression in the developing *Drosophila* eye. *Proc. Natl. Acad. Sci.* **86:** (in press).

Campos-Ortega, J.A., G. Jurgens, and A. Hofbauer. 1979. Cell clones and pattern formation: Studies on *seventess,* a mutant of *Drosophila melanogaster. Wilhelm Roux's Arch. Dev. Biol.* **186:** 27.

Hafen, E., K. Basler, J.E. Edstroem, and G.M. Rubin. 1987. *seventess,* a cell-specific homeotic gene of *Drosophila,* encodes a putative transmembrane receptor with a tyrosine kinase domain. *Science* **236:** 55.

Harris, W.A., W.S. Stark, and J.A. Walker. 1976. Genetic dissection of the photoreceptor system in the compound eye of *Drosophila melanogaster. J. Physiol.* **256:** 415.

Lawrence, P.A. and S.M. Green. 1979. Cell lineage in the developing retina of *Drosophila. Dev. Biol.* **71:** 142.

Ready, D.F., T.E. Hanson, and S. Benzer. 1976. Development of the *Drosophila* retina, a neurocrystalline lattice. *Dev. Biol.* **53:** 217.

Reinke, R. and S.L. Zipursky. 1988. Cell-cell interaction in the *Drosophila* retina: The *bride of sevenless* gene is required in photoreceptor cell R8 for R7 cell development. *Cell* **55:** 321.

Sibley, D.R., J.L. Benovic, M.G. Caron, and R.J. Lefkowitz. 1987. Regulation of transmembrane signaling by receptor phosphorylation. *Cell* **48:** 913.

Tomlinson, A. 1988. Cellular interactions in the developing *Drosophila* eye. *Development* **104:** 183.

Tomlinson, A., B.E. Kimmel, and G.M. Rubin. 1988. *rough,* a *Drosophila* homeobox gene required in photoreceptors R2 and R5 for inductive interactions in the developing eye. *Cell* **55:** 771.

Tomlinson, A., D.D.L. Bowtell, E. Hafen, and G.M. Rubin. 1987. Localization of the *sevenless* protein, a putative receptor for positional information in the eye imaginal disc of *Drosophila. Cell* **51:** 143.

von Heijne, G. 1986. Towards a comparative anatomy of N-terminal topogenic protein sequences. *J. Mol. Biol.* **189:** 239.

Developmental Potential of Embryonic Stem Cells

E.J. Robertson

Department of Genetics and Development, Columbia University
New York, New York 10032

In 1981, it was shown that permanent cultures of pluripotential embryonic stem cells (ES cells) could be obtained directly from outgrowths of preimplantation mouse embryos (Evans and Kaufman 1981; Martin 1981). Subsequently, it has been demonstrated that ES cells retain the ability to initiate an orderly and extensive program of differentiation when they are returned to the normal embryonic environment. Our studies have shown that ES cells integrate into the embryonic cell pool, differentiate in close association with the cells of the host carrier embryo, and routinely contribute extensively to both the embryonic and the extraembryonic cell lineages in a chimeric conceptus (Beddington and Robertson 1989). The pattern of colonization indicates that ES cells most closely resemble early inner cell mass cells in their developmental potential. Additionally, we have evaluated the behavior of a large number of independently derived ES cell lines in live-born chimeras and showed that the cells not only contribute to the somatic tissues, but also incorporated routinely into the germ line (for review, see Robertson and Bradley 1986). In particular, it is possible to maximize the efficiency of germ-line colonization through the use of cells carrying a functional Y chromosome as the incorporation of XY cells into XX host embryos causes a sex conversion effect to give male chimeras that transmit only ES-cell-derived sperm in the germ line.

Experimental Applications

Recent improvements in the techniques for introducing DNA into ES cells using $CaPO_4$ precipitation, electroporation, and retroviral vectors, together with the construction of both promoter elements and retroviral vectors, which express highly in embryo cells, afford unique possibilities for manipulating the mouse genome in culture. The use of ES cells as recipients for gene transfer experiments allows both studies of the transfected gene and any consequences of the incorporation of these cells into the tissues of a chimeric conceptus or adult to be monitored. ES cells have been used successfully to study the expression of the human type II collagen gene (Lovell-Badge et al. 1987) and the δ-crystallin gene (Takahashi et al. 1988), introduced into ES cells by transformation, in the developing mouse embryo. An advantage of this system is that the integration site, copy number, and expression can be characterized in cell clones prior to their introduction into the embryo. Additionally, the ES cell chimera system has a unique set of applications since candidate genes, placed in constructs suitably engineered to give inappropriate expression, can be transfected into ES cells. Although this form of genetic manipulation might be lethal in a transgenic embryo, generated by pronuclear injection, it will be entirely feasible to monitor any consequences of inappropriate expression within a chimeric conceptus.

One interesting application of the ES cell system is to introduce random mutations into the germ line after infection of cells in culture with a replication-defective retroviral vector (Robertson et al. 1986). Some of these proviral sequences may be inserted into genes that play an important role in development. The studies of Jaenisch and colleagues (for review, see Gridley et al. 1987), who used infectious virus as a means to generate and analyze insertional mutants, have highlighted this type of approach. Using ES cells, it is possible to increase substantially the overall number of insertional events by introducing multiple proviral sequences into the germ line, allowing a large number of individual inserts to be screened simultaneously in only a small number of animals. Preliminary analyses of these animals has identified a number of recessive mutations (E. Robertson, M. Kuehn, M. Evans, and A. Bradley, unpubl.).

Populations of ES cells can be mutagenized and screened in vitro to select for cells harboring specific genetic defects. Clones of XY ES cells selected and characterized in culture as carrying mutations in the hypoxanthine phosphoribosyltransferase (hprt) allele (leading to loss of function) have been used to construct chimeras. Transmission of HPRT-deficient cells in the germ line has allowed the establishment of three independent strains of mice that completely lack this enzyme activity and thereby provide a mouse model for the human X-linked genetic disease Lesch-Nyhan syndrome (Hooper et al. 1987; Kuehn et al. 1987). In light of the recent reports of efficient homologous recombination vectors to target mutations into cellular genes (for review, see Froman and Martin 1989), it is obvious that ES cells will become an increasingly important system whereby site-directed mutagenesis of specific loci will allow the role of developmentally regulated genes to be examined, as well as those implicated as having a causal role in disease.

Factors Affecting the Efficiency of Chimera Formation

Although it has now clearly been established that a wide variety of genetic manipulations can be successfully applied in tissue culture, the ultimate goal of such experiments is to achieve a germ-line transmission of the modified ES cell clones. It is therefore important to consider what factors will influence the developmental potential of a given ES cell clone. The most critical of these, with respect to germ-cell differentiation, is chromosome complement. Given that the majority of experimental strategies involve the selection and expansion of single-cell-derived colonies, it is obvious that the use of early passage cultures in which the majority of cells are euploid will maximize the chances of recovering normal cells.

A further consideration is whether the selection conditions will adversely affect the differentiation ability and hence chimera-forming ability of ES cells. We have tested the chimera-forming ability of the ES cell line CCE in experiments in which the cells were subjected to different experimental modifications. The results are summarized in Table 1. It is important to note that the culture conditions, choice of host blastocyst, and injection procedure are standardized. The parental cell line forms chimeras at a rate exceeding 95%, and this rate is not significantly depressed if the cell population is first infected with multiple copies of a replication-defective retroviral vector. Similarly, if the infected cells are selected for G418 resistance (G418^r), the chimera rate still exceeds 80%. It should also be noted that ES cell clones derived from the D3 cell line, transfected with pSV2*neo* plasmid, and selected for resistance to G418, are also capable of colonizing the germ line (Gossler et al. 1986). Table 1 shows that two different clonal derivatives, selected for resistance to 6-thioguanine (6-TG) (resulting from loss of function of the *hprt* enzyme), both gave a lower rate of chimera formation when compared with the parental culture (42% and 75%, respectively). Although, for both clones, germ-line transmission of the TG-resistant cells was obtained (indicating that they retained a normal karyo-

41

Table 1 Characteristics of Chimeras Generated Using the CCE ES Cell Line

Cell culture	No. embryos injected	No. chimeras/ no. born	No. females	No. males	Fertile males	Germ-line males
[a]CCE parental	33	17/17	2	15	13	5
[a]CCE virally infected	170	89/114	18	64	56	23
[b]CCE G418[r]	57	24/28	5	19	5	2
[c]CCE-TG3	266	66/157	16	50	36	3
[c]CCE-TG4	228	94/143	42	52	27	7

Chimeras were obtained by injecting between 10–15 individual XY CCE ES cells into blastocysts derived from the MFI albino outbred mouse strain. Chimeras were scored on the basis of coat color (CCE is derived from the 129/Sv inbred strain and is homozygous for the black agouti coat color genes). To test for germ-line contribution, male chimeras were test bred with albino females, and the resulting litters were scored for the presence of pigmented offspring.

[a]Data from Robertson et al. (1986) and E. Robertson and A. Bradley (unpubl.). The retroviral vector and infection protocols used are described in Robertson et al. (1986).

[b]A. Bradley and E. Robertson (unpubl.). Selection used was 0.3 mg/ml G418.

[c]Data from Kuehn et al. (1987). Note that the sex ratio of CCE-TG4 chimeras is not distorted, since the clonal line is composed of both XO and XY cells. Breeding data from chimeric females is not included.

type), it is interesting to note that the relative contribution by the CCE-TG3 and -TG4 to the somatic tissues of chimeras was less extensive than that seen in parental CCE chimeras (data not shown). Two other reports of germ-line transmission of 6-TG- or HAT-selected ES cell clones derived from the E14 cell line (Hooper et al. 1987; Thompson et al. 1989) provide additional evidence that this form of selection does not compromise the differentiation ability of ES cells.

Finally, there is some evidence that the genetic background of the ES cell and the host blastocyst may be important. It has been well documented from embryo aggregation chimeras that strain effects exist. For example, Petersen et al. (1979) showed that in 129<—>C57BL/6 chimeras, the 129 genotype predominates in all tissues examined. Our experiments using 129/Sv-derived ES cells show that they will also contribute substantially to the somatic tissues and germ line in combination with outbred MF1 strain embryos. This contrasts with the results of Suda et al. (1987) using the 129/Sv-derived CPI line in combination with embryos derived from the outbred albino CD1 strain.

In summary, it seems that a number of factors may influence the developmental ability of ES cells, and that these, in turn, will determine the difficulty of transferring mutations introduced in culture back into the mouse germ line. These may include more obvious factors, such as culture protocols and selection procedures, in addition to more subtle effects relating to genetic differences between ES cells and the embryonic environment to which they are returned.

REFERENCES

Beddington, R.S.P. and E.J. Robertson. 1989. An assessment of the developmental potential of embryonic stem cells in the developing embryo. *Development* **105**: 733.

Evans, M.J. and M.H. Kaufman. 1981. Establishment in culture of pluripotential cells from mouse embryos. *Nature* **292**: 154.

Froman, M.A. and G.R. Martin. 1989. Cut, paste, and save: New approaches to altering specific genes in mice. *Cell* **56**: 145.

Gossler, A., T. Doetschman, R. Korh, E. Serfling, and R. Kemler. 1986. Transgenesis by means of blastocyst-derived embryonic stem cell lines. *Proc. Natl. Acad. Sci.* **83**: 9065.

Gridley, T., P. Soriano, and R. Jaenisch. 1987. Insertional mutagenesis in mice. *Trends Genet.* **3**: 162.

Hooper, M., K. Hardy, A. Handyside, S. Hunter, and M. Monk. 1987. HPRT-deficient (Lesch-Nyhan) mouse embryos derived from germ line colonization by cultured cells. *Nature* **326**: 292.

Kuehn, M.R., A. Bradley, E.J. Robertson, and M.J. Evans. 1987. A potential animal model for Lesch-Nyhan syndrome through introduction of HPRT mutations into mice. *Nature* **362**: 295.

Lovell-Badge, R.H., A. Bygrave, A. Bradley, E. Robertson, R. Tilly, and K.S.E. Cheah. 1987. Tissue specific expression of the human type II collagen gene in mice. *Proc. Natl. Acad. Sci.* **84**: 2803.

Martin, G.R. 1981. Isolation of a pluripotential cell line from early mouse embryos cultured in medium conditioned with teratocarcinoma cells. *Proc. Natl. Acad. Sci.* **78**: 7634.

Peterson, A.C., P.M. Frair, H.R. Rayburn, and D.P. Cross. 1979. Development and disease in the neuromuscular system of muscular dystrophic <—> normal mouse chimeras. *Soc. Neurosci. Symp.* **4**: 258.

Robertson, E.J. and A. Bradley. 1986. Production of permanent cell lines from early embryos and their use in studying developmental problems. In *Experimental approaches to mammalian embryonic development* (ed. J. Rossant and R.A. Petersen), p. 475. Cambridge University Press.

Robertson, E., A. Bradley, M. Kuehn, and M.J. Evans. 1986. Germline transmission of genes introduced into cultured pluripotential cells by a retroviral vector. *Nature* **323:** 445.

Suda, Y., M. Suzuki, Y. Ikawa, and S. Aizawa. 1987. Mouse embryonic stem cells exhibit indefinite proliferative potential. *J. Cell. Physiol.* **133:** 197.

Takahashi, Y., K. Hanaoka, M. Hayasaka, K. Katoh, Y. Kato, T.S. Okada, and H. Kondoh. 1988. Embryonic stem cell-mediated transfer and correct regulation of the chicken δ-crystallin gene in developing mouse embryos. *Development* **102:** 259.

Thompson, S., A.R. Clarke, A.M. Pow, M.L. Hooper, and D.W. Melton. 1989. Germ line transmission and expression of a corrected HPRT gene produced by gene targeting in embryonic stem cells. *Cell* **56:** 313.

Creating Mice with Specific Mutations by Gene Targeting

M.R. Capecchi, K.R. Thomas, and S.L. Mansour

Department of Biology, Howard Hughes Medical Institute
University of Utah, Salt Lake City, Utah 84112

Homologous recombination between DNA sequences residing in the chromosome and newly introduced cloned DNA sequences (i.e., gene targeting) allow the transfer of any modification of the cloned gene into the genome of a living cell (for review, see Capecchi 1989a,b). Furthermore, if the recipient cell is a pluripotent, mouse-embryo-derived stem (ES) cell, it is possible to transfer that modification, created in a test tube, to the germ line of a living mouse. Thus, the potential now exists for modifying, in a defined manner, any cloned mouse gene and then evaluating the phenotypic consequences of that modification on the mouse.

The ability to generate specific mouse mutations via gene targeting should have a major impact on many phases of mammalian biology including development, cancer, immunology, neurobiology, and human medicine. In the context of this volume, recent molecular genetic analysis of development in *Drosophila* has revealed a network of genes that controls the formation of its metameric pattern (for review, see Ingham 1988). Many of these genes have been found to contain a common DNA sequence domain, the homeo box, that encodes a DNA-binding domain (McGinnis et al. 1984; Scott and Weiner 1984). On the basis of DNA sequence similarity, related genes, specifically the *Hox* genes, have been identified in the mouse (for review, see Dressler and Gruss 1988; Holland and Hogan 1988). The function of these genes in the mouse is not known. However, based on the conservation of the genome organization of the *Hox* genes relative to the *Drosophila* genes, as well as on their embryonic expression patterns (see Kappen et al.; Galliot et al.; Graham et al.; all this volume), it is safe to postulate that these genes are required for establishing positional information during development in the mouse.

To assign a functional role to some of these genes, we are in the process of generating null alleles of homeo-box-containing genes belonging to the *Hox-1* and *Hox-2* linkage groups. Targeted disruption of these genes should not only reveal the phenotypes associated with inactivation of the individual genes, but, through epistatis and molecular analysis, may also help define how these genes participate in the developmental network controlling early mouse organogenesis and morphogenesis.

In addition, we have initiated a genetic analysis of genes participating in developmental decisions through cell-cell interactions. Specifically, we are concentrating on the *int*-related genes. This functionally related family of genes was first identified as genes activated in mammary tumors of mice by the nearby insertion of the mouse mammary tumor virus (for review, see Nusse 1988). Subsequently, it has been shown that the protein products of these genes are secreted and often exhibit sequence similarities to growth factors. In situ hybridization analysis of the mouse reveals diverse but highly restricted patterns of expression during development. In the case of *int-1*, a *Drosophila* homolog, *wingless* (*wg*), has been identified that belongs to the segment polarity class of genes. *wg* appears to be required to specify the identity of cells within a segment (see Nusse et al., this volume).

From ES Cells to Germ-line Chimera
Figure 1 outlines the procedure for introducing a desired mutation into the germ line of mice using targeted modification of the ES cell genome as an intermediate. A targeting vector, containing the desired mutation, is introduced into ES cells by electroporation or microinjection. In most cells, the targeting vector inserts randomly into the ES genome. However, in a few cells, the targeting vector pairs with the cognate chromosomal DNA sequence and transfers the mutation to the genome by homologous recombination. Screening and/or enrichment procedures are then used to identify the rare ES cell in which the targeting event has occurred. The appropriate cell is then cloned and maintained as a pure population. Next, the altered ES cells are injected into the blastocoel cavity of a preimplantation mouse embryo and the blastocyst is surgically transferred into the uterus of a foster mother where development is allowed to progress to term (Bradley et al. 1984). The resulting animal is chimeric in that it is composed of cells derived from

both the donor stem cells and the host blastocyst. In the particular example shown in Figure 1, the ES cells are derived from a mouse homozygous for the black coat color allele, and the recipient blastocyst is derived from an albino mouse. The resulting chimeric mouse is composed of cells of both genotypes and therefore displays a coat with patches of each color. Breeding of the chimeric mouse to an albino mouse yields some black mice, indicating that the ES cells contributed to the formation of the germ line. Genomic screening of these progeny is used to determine which mice received the allele carrying the targeted mutation. Interbreeding of heterozygous siblings yields animals homozygous for the desired mutation.

Disruption of the *hprt* Gene

Mammalian cells can mediate recombination between homologous DNA sequences, but they demonstrate an even greater propensity for mediating nonhomologous recombination. The problem is thus to identify homologous recombination events in a vast pool of scattered, nonhomologous recombination events. In this context, the hypoxanthine phosphoribosyltransferase (*hprt*) gene has provided an ideal model system for developing the technique of gene targeting in ES cells because one can select directly for the targeting event. Since this gene is on the X chromosome, only one mutant copy is needed to yield the recessive *hprt⁻* phenotype in male ES cells. The *hprt⁻* cells are selected by growth in the presence of the base analog, 6-thioguanine (6-TG), which kills *hprt⁺* cells.

Two classes of targeting vectors were tested for their ability to disrupt the *hprt* gene (Thomas and Capecchi 1987): sequence replacement and sequence insertion vectors. Using yeast as a paradigm, we anticipated that sequence replacement vectors would replace endogenous DNA with exogenous sequences, whereas sequence insertion vectors would insert the entire vector DNA sequence into the endogenous locus. Each class of vectors contains a neomycin resistance gene within the eighth exon of *hprt*. This arrangement not only disrupts the coding sequence of *hprt*, but also provides a selectable marker for cells containing an integrated copy of the recombinant vector (resistance to the drug G418).

After the introduction of the targeting vectors into ES cells by electroporation and the selection for resistance to G418 and 6-TG, all survivors were found to have lost *hprt* activity as a result of targeted disruption of the *hprt* gene. Interestingly, re-

47

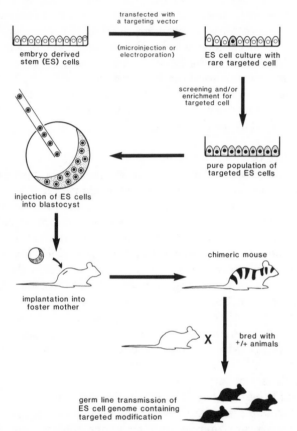

Figure 1 Generation of mouse germ-line chimeras from ES cells containing a targeted gene disruption. A targeting vector, containing the desired mutation, is introduced into ES cells by microinjection or electroporation. A screening and/or enrichment procedure is then used to identify the rare ES cells in which the targeting event occurred. Appropriate cells are cloned and injected into a recipient blastocyst. The blastocyst is then transferred into the uterus of a foster mother. The resulting animal is chimeric (i.e., composed of cells derived from both the donor stem cell and the recipient blastocyst). ES cells were derived from a mouse homozygous for the black coat color allele and the recipient blastocyst from a mouse homozygous for the albino allele. Chimeric mice therefore display mosaic black and white coats. Breeding of a chimeric mouse to an albino mouse yielded some black mice from which we can conclude that the ES cells contributed to the formation of the germ line.

48

placement vectors and insertion vectors were equally efficient at disrupting the endogenous *hprt* gene. Furthermore, both vectors showed the same strong dependency of the targeting frequency on the extent of homology between the targeting vector and the endogenous DNA sequence. Over the range tested, from 2.9 to 14.3 kb, a 5-fold increase in DNA sequence homology resulted in roughly a 100-fold increase in the targeting frequency (Capecchi 1989b).

Nonselectable Genes

The *hprt* gene was chosen as a model in the initial studies because direct selection could be used to isolate cells in which a homologous recombination event had occurred and because it is present as only a single copy in male cells. However, most genes are present as two copies in the genome, and in the vast majority of cases a selectable cellular phenotype is not associated even with the inactivation of both copies, let alone with the more frequent single-copy inactivation event. Therefore, it is desirable to have some means for identifying the rare ES cell in which a nonselectable gene has been inactivated; this identification can be achieved by using an indirect enrichment and/or screening procedure.

We have chosen to pursue developing enrichment procedures rather than screening procedures for identifying ES cells in which a targeting event occurred because, if successful, they should be less labor intensive and should permit the identification of rarer events. At present, it is difficult to predict how much the targeting frequency will vary from locus to locus. In particular, the frequency of disrupting a gene that is not expressed in ES cells remains unknown.

Recently, we have described an enrichment procedure that is independent of the function of the target gene (Mansour et al. 1988). This procedure uses a positive selection for cells that have incorporated the targeting vector anywhere in the ES genome and a negative selection against cells that have randomly integrated the vector. The net effect is to enrich for cells containing the desired targeted mutation. We have termed this enrichment procedure the positive-negative selection (PNS) procedure.

PNS was used to enrich for ES cells containing disruptions of the *hprt* and *int-2* genes. Following introduction of the *hprt*-PNS-targeting vector into ES cells, virtually all (19 of 24) of the

49

selected colonies contained disruption of the *hprt* gene even though we did not select for the *hprt*⁻ phenotype. The frequency of successful targeting to the *int-2* locus among PNS-selected cells was 20-fold lower; in 20 cell lines, 1 contained a disrupted *int-2* gene. This lower frequency may reflect the lower level of *int-2* expression in ES cells. *hprt* transcripts can be readily detected in ES cells, whereas *int-2* is expressed at a level of less than one transcript per cell. The PNS procedure has also been used to create null mutations in the mouse homeo-box-containing genes *Hox-1.2* and *Hox-1.3* (D. Kostic and M. Capecchi, unpubl.). Like *hprt*, *Hox-1.2* and *Hox-1.3* are expressed at moderate levels in ES cells and exhibited gene-targeting disruption frequencies similar to those obtained with *hprt*.

An ideal strategy for disrupting a gene not expressed in ES cells has not yet emerged. Among alternative approaches to consider are the use of pure screening procedures or enrichment procedures. Embedded within the decision of whether to use nonselective or selective protocols is the choice of whether to deliver the targeting vector to the recipient cells by micro-injection or by electroporation. Recently, we have been successful in using the PNS procedure to disrupt the mouse *int-1* gene. Using methodology that can detect *int-2* transcripts in ES cells, we could not detect *int-1* transcripts. Thus, *int-1* is transcribed in ES cells at extremely low levels, if at all. The targeting frequency for disrupting the *int-1* gene was approximately 100-fold lower than observed for *hprt*. We are currently working on procedures to increase this frequency.

The Future

As stated, we are using gene targeting to disrupt genes belonging to two classes: those encoding transcription factors that activate the progression of the developmental program of the mouse and those that mediate cell-cell interactions, which at a local level presumably feed back information to the program concerning developmental progress and thereby direct regional decisions. *Drosophila* has been used as a paradigm to identify mouse genes belonging to the first class (i.e., the homeo-box-containing genes). Implicit in this operation is the hope that the same transcription factors that control the formation of the metameric pattern of *Drosophila* and the identity of cells within segments will mediate equally important, if not identical, decisions concerning positional information within the developing mouse embryo. Hopefully, generating mice with null alleles in these homeo-box-containing genes will yield informative phenotypes.

The *int*-related genes provide an attractive set of genes for developmental involvement. They exhibit restricted but different patterns of gene expression during development. As discussed above, a *Drosophila* homolog for *int-1* has been identified. Clonal analysis of this mutant demonstrated that *wg⁻* cells can be rescued by surrounding wild-type cells, in keeping with the hypothesis that the *wg* product is exported by one set of cells and is affecting the identity of neighboring cells.

No *Drosophila* homolog for *int-2* has been identified. In fact, this gene may be restricted to mammals. However, *int-2* protein does appear to be secreted and exhibits protein sequence similarity to members of the fibroblast growth factor family. Expression studies in the mouse and transplantation experiments in chickens suggest a role for *int-2* as an embryonic inducer during the formation of the otocyst (i.e., a hollow chamber of ectoderm that develops into the inner ear).

The efficacy of using targeted ES cells to create mice with specific mutations is dependent on the ability to use the modified ES cells to generate germ-line chimeras. To date, Thompson et al. (1989) have reported making germ-line chimeras from ES cells in which a mutant *hprt* gene had been corrected by gene targeting. A number of laboratories have generated chimeric mice from ES cells harboring targeted mutations in genes of developmental interest and are in the process of evaluating germ-line transmission. A variety of strategies are being tested to determine factors that may influence germ-line transmission. The most promising among these approaches is the evaluation of the compatibility of the recipient blastocyst derived from different inbred and outbred mouse strains when paired with specific ES cell lines.

REFERENCES

Bradley, A., M. Evans, M.H. Kaufman, and E. Robertson. 1984. Formation of germ-line chimeras from embryo-derived teratocarcinoma cell lines. *Nature* **309:** 255.

Capecchi, M.R. 1989a. The new mouse genetics: Altering the genome by gene targeting. *Trends Genet.* **5(3):** 72.

———. 1989b. Altering the genome by homologous recombination. *Science* **244:** 1288.

Dressler, G.R. and P. Gruss. 1988. Do multigene families regulate vertebrate development? *Trends Genet.* **4(8):** 214.

Holland, P.W.M. and B.L.M. Hogan. 1988. Expression of homeo box genes during mouse development: A review. *Genes Dev.* **2:** 773.

Ingham, P.W. 1988. The molecular genetics of embryonic pattern formation in *Drosophila*. *Nature* **335:** 25.

Mansour, S.L., K.R. Thomas, and M.R. Capecchi. 1988. Disruption of the proto-oncogene *int-2* in mouse embryo-derived stem cells: A general strategy for targeting mutations to non-selectable genes. *Nature* **336:** 348.

McGinnis, W., M.S. Levine, E. Hafen, A. Kuroiwa and W.J. Gehring. 1984. A conserved DNA sequence in homeotic genes of the *Drosophila* Antennapedia and Bithorax complexes. *Nature* **308:** 428.

Nusse, R. 1988. The *int* genes in mammary tumorigenesis and in normal development. *Trends Genet.* **4(10):** 291.

Scott, M.P. and A.J. Weiner. 1984. Structural relationships among genes that control development: Sequence homology between the Antennapedia, Ultrabithorax, and fushi tarazu loci of *Drosophila*. *Proc. Natl. Acad. Sci.* **81:** 4115.

Thomas, K.R. and M.R. Capecchi. 1987. Site-directed mutagenesis by gene targeting in mouse embryo-derived stem cells. *Cell* **51:** 503.

Thompson, S., A.R. Clark, A.M. Pow, M. Hopper, and D.W. Melton. 1989. Germline transmission and expression of a corrected *hprt* gene, produced by gene targeting in embryonic stem cells. *Cell* **56:** 313.

Precision Mutagenesis by Homologous Recombination

A. Zimmer and P. Gruss

Max-Planck-Institute of Biophysical Chemistry, Department of
Molecular Cell Biology, 3400 Göttingen, Federal Republic of Germany

Understanding the mechanisms that control the differentiation
of the fertilized egg into the various cell types of the adult
organism is one of the major goals of embryology. In lower
organisms, this goal has been approached by a comprehensive
genetic analysis. In higher organisms, like the mouse, such an
analysis turned out to be unfeasible. Here reverse genetics
must substitute the classical strategies.

Putative Mammalian Developmental Control Genes: Cloning with *Drosophila* Probes

Despite the great evolutionary distance that separates insects
and vertebrates, both groups of organisms have highly
homologous genes, which have been shown to play a key role at
least in *Drosophila* development. This homology became first
evident when mammalian genes containing a homeo box were
cloned. The homeo box is a 180-bp sequence originally identified as a conserved element in the *Drosophila* segmentation
and homeotic genes *Antennapedia (Antp), Ultrabithorax,* and
fushi tarazu. It encodes for a sequence-specific DNA-binding
domain termed the homeo domain. The DNA binding is
mediated by a part of the homeo domain that folds into a helix-
turn-helix structure, similar to that found in DNA-binding
proteins of *Saccharomyces cerevisiae (MATα1* and *MATα2)* or
even prokaryotic DNA-binding proteins (λ-repressor, cap, and
cax) (for review, see Gehring and Hiromi 1986). Using the
Drosophila homeo box as a molecular probe, at least 23 murine
genes with a highly homologous homeo box have been isolated.
They are clustered on chromosomes 2, 6, 11, and 15 (for review,
see Dressler and Gruss 1988). In addition to the *Antp*-type
homeo box, different classes of homeo boxes have been found in
several other *Drosophila* developmental genes, such as
*engrailed (en), even-skipped (eve), paired (prd), bicoid, caudal
(cdx),* and *muscle segment homeo box* gene. Again, a systematic

screening revealed that the murine genome also contains homeo box genes that belong to these classes. Examples are the murine *en-1; en-2; cdx; paired box (Pax 1, 3,* and *7); Hox-7;* and *eve 1* and *2* (for review, see Akam 1987; Zimmer and Gruss 1989b; H. Bastian, pers. comm.). The expression of the murine genes has been analyzed in detail during mouse embryogenesis: All murine homeo box genes of the *Antp*-type are expressed in a temporally and spatially restricted pattern in the central and peripheral nervous system, in the neural crest and neural-crest-derived tissues, in mesoderm-derived somites, and in visceral organs and the intestine (for review, see Holland and Hogan 1988). This expression pattern is compatible with the assumption that murine homeo-box-containing genes are involved in the determination of regional specificity.

It was not surprising that using another DNA-binding motif present in *Drosophila* and *Xenopus* genes, the zinc finger motif, numerous murine homologous genes could also be isolated (Chowdhury et al. 1987). Finally, a conserved element of a yet unknown function called the paired domain was identified in the *Drosophila prd, gooseberry-proximal,* and *gooseberry-distal* genes and has been used to isolate murine homologs: the *prd* gene family (for review, see Dressler and Gruss 1988; Zimmer and Gruss 1989b).

First Functional Assignments: The *Pax 1* Gene
Obviously, the question arose whether one of these isolated genes could be correlated with one of the known mouse developmental mutants. To do so, the chromosomal location of these genes was determined, and it was compared with the location of known mutations. So far, only a few candidates were available. *Hypodactyly,* a mutation affecting the limb bud formation has been mapped within 1 cM of the *Hox-1.1* cluster. *Luxate, Hemimelic extratoe,* and *Hammertoe,* all mutations that involve changes in the pattern formation in limbs map close to *Hox-7,* a gene expressed in limb buds. However, none of these mutants could be linked with a gene function. Thus, it was exciting that such an assignment could be made between the *Pax 1* gene and the *undulated (un)* mutation. *Pax 1* is expressed from day 9 postcoitus to day 15 postcoitus along the rostral-caudal axis in the developing embryo. It is first detected in the ventral mesenchyme lateral to the notochord, later in perichordal condensations around the notochord, and finally in the intervertebral disks. The *un* mice exhibit malformations in the entire axial

skeleton with enlarged intervertebral disks and smaller verte-
bral centers. In addition, *Pax 1* expression was found in the
sternum, which is also affected in the *un* mutation. Balling et
al. (1988) demonstrated that the *Pax 1* gene maps to the same
chromosomal region as *un*. In *un* mice, *Pax 1* has an amino acid
substitution in the paired box. This mutation was not found in
any of the other analyzed mouse strains, thus providing com-
pelling evidence that it causes the *un* phenotype.

The Next Step: Specific Mutations Can Be Introduced into Mice

Our ability to analyze the function of homeo box, paired box, or
finger genes no longer depends on serendipity. We recently described a technique that allows the introduction of specific
mutations into mice. This technique involves the introduction
of a chosen mutation by standard recombinant DNA technology
into a cloned genomic sequence. This mutation is then transferred into embryonic stem (ES) cells. ES cell lines are derived
from the inner cell mass of mouse blastocysts. They are
pluripotent and can be handled and manipulated like any other
cell line. When reintroduced into mouse blastocysts, they participate in the normal development and contribute to all tissues, including the germ cells. Thus, chimeras that are generated with mutated ES cells can be bred to yield heterozygous
and eventually homozygous mutant mice.

We have used the *Hox-1.1* gene as a model gene to establish
the technique (Zimmer and Gruss 1989a). Earlier gene-
targeting experiments used a neo-resistant gene driven by a
promoter/enhancer to disrupt the coding sequence of the target
gene (Thomas and Capecchi 1987). We were concerned about
the introduction of a new enhancer into the *Hox* cluster that
might interfere with the expression of adjacent *Hox* genes during the mouse development being detrimental for the analysis
of the mutant phenotype. Thus, we decided to avoid the need
for selection by the use of microinjection, the most efficient
method for gene transfer, and the highly sensitive polymerase
chain reaction (PCR) in the analysis of the recipient cell pools.
We used this method to insert a 20-bp oligonucleotide into the
homeo box of the *Hox-1.1* gene. This oligonucleotide introduced
a frame shift into the *Hox-1.1* gene and served as a priming site
in the PCR analysis (see Fig. 1). A 1.56-kb fragment containing
the intron and the second exon with the mutation was microinjected into pools of ES cells. After expansion of the pools,

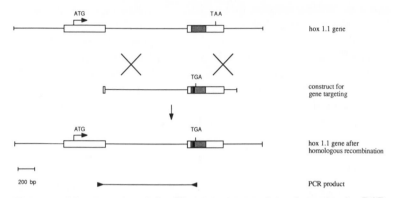

Figure 1 Mutagenesis of the *Hox-1.1* gene, and its detection by PCR. The intron and the second exon of the gene were cloned, and a 20-bp oligonucleotide was inserted into the homeo box. Homologous recombination of the endogenous *Hox-1.1* gene with this construct introduced this oligonucleotide into the endogenous gene. These events were detected by PCR. One primer in this reaction was identical to the inserted oligonucleotide. The other primer was specific for the first exon of the *Hox-1.1* gene outside of the cloned region. Thus, only a mutated allele contains both priming sites and can give rise to an amplification of the indicative fragment.

DNA was analyzed by PCR. We found that 5 of the 12 microinjected pools contained homologously recombined cells. We calculated that 1 of 150 microinjected cells had undergone homologous recombination. This was surprising because comparable studies revealed a much lower frequency of homologous recombination. To test how the length of the homologous sequences influence the frequency, we injected two different constructs, all harboring the same 20-bp oligonucleotide insertion, into P19 teratocarcinoma cells. Fragment II has an extended homology with the 3′ end and a total homologous region of 4 kb. Fragment III extends further in the 3′ direction for a total homology of 8.5 kb. A PCR analysis of these experiments is shown in Table 1. We found that the frequency of homologous recombination was still increased with fragment II to yield one homologous recombination event in 130 injected cells. Doubling the homologous region with fragment III will further increase this frequency to 1 in 110 injected cells (see Table 1).

So far, at least ten chimeric males have been obtained with the mutated ES cells. Five have been tested for germ-line transmission of the mutated allele. None of them produced offspring derived from the mutated ES cells. The other animals

Table 1 PCR Analysis of Microinjected P19 Teratocarcinoma Cells

Fragment II				Fragment III		
exp.	no. of cells	PCR		exp.	no. of cells	PCR
1	50	–		1	48	–
2	50	+		2	50	+
3	50	+		3	50	+
4	50	–		4	50	+
5	50	+		5	50	+
6	50	+		6	35	–
7	50	–		7	40	–
8	50	–		8	50	+
				9	50	–
				10	50	–
				11	37	–
				12	22	–
				13	50	+
				14	38	–
				15	49	–

are currently set up for breeding. Although we cannot exclude the possibility that the mutation prevents germ-line transmission, we feel that our results most likely reflect the initial difficulties in setting up the technique. Thus, more chimeras must be generated, and different batches of ES cells must be used to guarantee the final success.

REFERENCES

Akam, M. 1987. The molecular basis for metameric pattern in the *Drosophila* embryo. *Development* **101:** 1.

Balling, R., U. Deutsch, and P. Gruss. 1988. *undulated*, a mutation affecting the development of the mouse skeleton, has a point mutation in the paired box of *Pax 1*. *Cell* **55:** 531.

Chowdhury, K., U. Deutsch, and P. Gruss. 1987. A multigene family encoding several "finger" structures is present and differentially active in mammalian genomes. *Cell* **48:** 771.

Dressler, G.R. and P. Gruss. 1988. Do multigene families regulate vertebrate development? *Trends Genet.* **4:** 214.

Gehring, W.J. and Y. Hiromi. 1986. Homeotic genes and the homeobox. *Annu. Rev. Genet.* **20:** 147.

Holland, P.W.H. and B.L.M. Hogan. 1988. Expression of homeo box genes during mouse development: A review. *Genes Dev.* **2:** 773.

Thomas, K.R. and M.R. Capecchi. 1987. Site-directed mutagenesis by gene targeting in mouse embryo-derived stem cells. *Cell* **51:** 503.

Zimmer, A. and P. Gruss. 1989a. Production of chimeric mice containing embryonic stem (ES) cells carrying a homeobox *Hox 1.1* allele
mutated by homologous recombination. *Nature* **338:** 150.
―――. 1989b. New strategies in developmental biology: In vivo
mutagenesis as a tool to dissect mammalian development. *NATO
ASI Ser. Ser. H. Cell Biol.* (in press).

Molecular Genetic Approaches to the Analysis of Mammalian Development

A.L. Joyner, C. Davis, C. Moens, and J. Rossant

Division of Molecular and Developmental Biology, Mt. Sinai Hospital
Research Institute and Department of Medical Genetics
University of Toronto, Toronto, Ontario M5G 1X5

Recent advances in molecular genetic techniques and in our ability to manipulate the mouse germ line have offered new opportunities for studying the genetic control of mammalian development where alternative approaches such as standard mutational analysis have proven difficult to carry out. We are particularly interested in characterizing the genes that control pattern formation in mammals with the aim of identifying genes that act sequentially to lay down the basic body plan. The limited experimental evidence available on mouse development would suggest that these genes act during, or just prior to, gastrulation when the cells of the embryo begin to rapidly proliferate and migrate. This cell activity appears to be accompanied by a progressive loss of developmental potential and/or the acquisition of specialized fates. The first segregation of cells is into one of the three germ layers, followed by differentiation of specific tissues, such as the neural tube and somites, within each germ layer. Of interest is how groups of cells within these tissues then become differentially specialized along the body axis. We would like to identify the genetic or environmental cues responsible for this seemingly position-dependent specialization, as well as the genes that respond to these cues and presumably direct the cells down different developmental pathways.

The identification of conserved protein domains among many *Drosophila* pattern-formation genes and their presence in higher organisms has suggested one possible route for identifying developmental genes in mammals (McGinnis et al. 1984). We have used one such conserved protein domain found in the *Drosophila engrailed (en)* and *invected* genes, termed the *engrailed*-conserved region and including the homeo box and 60 amino acids of adjacent sequences, to clone out two genes, *En-1* and *En-2*, from the mouse (Joyner et al. 1985; Joyner and Martin 1987), human (Logan et al. 1989), and chicken (D. Nallainathan and A. Joyner, unpubl.). Starting with these candidate developmental gene sequences, we have begun to determine the function of their respective genetic loci by combining molecular genetic techniques, such as in situ analysis of expression, with a mutational analysis using homologous recombination in embryonic stem cells to generate mutant mice.

Molecular Analysis of *En-2*

To take a molecular genetic approach to studying the function and regulation of developmental genes, the genomic structure of the genes must be determined and their protein products identified. The mouse *En-2* gene transcribes one major 3.7-kb mRNA (Joyner and Martin 1987; Davis et al. 1988), and a 3-kb cDNA clone representing its 3′ end has been cloned (Joyner and Martin 1987). We have recently sequenced this cDNA clone and 1 kb of genomic DNA upstream of the 5′ end of the cDNA, as well as the corresponding human *EN2* genomic DNA sequences (C. Logan and S. Noble-Topham, unpubl.). The results of this analysis predict that a single mRNA is transcribed from the mouse and human *En-2* genes that can code for a protein with an approximately 38-kD molecular mass.

To allow us to begin to analyze the *En* protein products, polyclonal rabbit antiserum was raised to a bacterial *TrpE-En-2* fusion protein that contained the *engrailed*-conserved region of *En-2*. On the basis of Western blot analysis of tissues that express each of the *En* genes, the affinity-purified serum appears to be specific for the *En* genes and detects an approximately 40-kD *En-2* and 50-kD *En-1* protein. This serum will now be used to examine whether the *En* proteins are posttranslationally modified, as is the *Drosophila en* protein (Gay et al. 1988).

Expression of the *En* Genes during Development

In situ RNA hybridization analysis is a very powerful and relatively straightforward first approach that can be taken, once a candidate developmental gene has been cloned, to obtain initial insight into the possible function of a gene in development. We have used in situ RNA hybridization with *En-1-* and *En-2-* specific probes to analyze the expression of each gene throughout most of development and in the adult brain (Davis and Joyner 1988; Davis et al. 1988). Expression of both genes was

60

first detected in seven somite embryos in nearly identical bands of cells across the neural folds and remained as homogenous bands or rings of cells around the neural tube at the midhind brain junction until neurogenesis occurred. At this point, expression of each gene appeared to become more and more restricted to overlapping but different specific groups of cells in the cerebellum and pons region. In addition, *En-1* expression was found to be dramatically different from *En-2* outside the brain during midgestation. Whereas *En-2* seemed to be completely restricted in the brain, *En-1* was also expressed at 12.5 days in two somite-derived tissues, in two ventral-lateral stripes in the spinal cord, and in the limb and tail buds.

We have recently begun to use the *En*-affinity-purified antisera to analyze *En* protein expression in mouse and chicken embryos. One advantage of using antibodies to detect expression patterns is that early embryos can be analyzed in whole mount, and thus the overall three-dimensional pattern can be easily visualized. The pattern of *En* protein expression detected in 8.5–11-day-old whole mouse embryos was identical with that previously detected for *En-1* and *En-2* by RNA in situ hybridization analysis (Davidson et al. 1988; Davis and Joyner 1988; Davis et al. 1988).

We are also interested in studying the *En* genes in the chicken because chicken embryos offer a number of experimental advantages over mice. Our preliminary analysis of *En* expression in chicken embyros (stages 8–17) using the purified *En* sera showed that expression in the brain was very similar to that in the mouse and to that seen previously in chickens using *En-2* probes (Gardner et al. 1988). The *En* antisera also detected expression in the somites and possibly the spinal cord and limb buds that was very similar to that seen for *En-1* in the mouse (K. Millen and A. Joyner, unpubl.).

Overall, expression of the *En* genes first appears to mark spatial domains within a number of tissues prior to their terminal differentiation and later only a subset of differentiated cells within these tissues. These genes may therefore function both in pattern formation and terminal differentiation within specific tissues.

Production of Mutations in Developmental Genes

The molecular genetic approaches described above and applied to the *En* genes, although giving us some insights into the potential roles of these genes in development, are not sufficient to determine their function. An analysis of the phenotypic effect of mutations in developmental genes may address this problem more directly. To this end, we have begun to develop vectors and methods of screening for targeted mutations following homologous recombination in embryonic stem (ES) cells. We have obtained three ES cell lines in which one allele of *En-2* is mutated (Joyner et al. 1989). The vector used contained *En-2* genomic sequences in which part of the homeo-box-containing exon was replaced by the selectable gene *neo* driven by the human β-actin promoter and lacking a polyadenylation site. To screen for targeted integrations, we used the polymerase chain reaction (PCR) to screen pools of *neo*-resistant (*neo*[r]) colonies for a novel *neo/En-2* junction fragment. The frequency of homologous recombination obtained was 1 in 260 *neo*[r] colonies.

More recently, we have obtained cell lines carrying a mutation in one allele of N-*myc*. For these experiments, we took advantage of the fact that N-*myc* is expressed at high levels in ES cells. The vector used for these experiments is shown in Figure 1, and following homologous recombination, results in the production of an N-*myc/neo* fusion transcript in which expression of *neo* is under the control of the N-*myc* promoter/enhancer. By

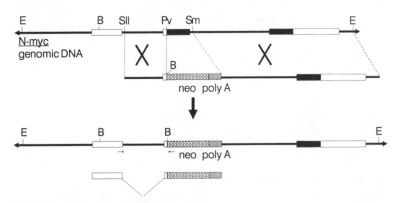

Figure 1 Homologous recombination scheme for N-*myc*. Schematic diagram showing the N-*myc* locus (*top*), the targeting vector (*upper middle*), the predicted sequences present in N-*myc* following homologous recombination of the vector (*lower middle*), and the structure of the predicted N-*myc*/neo fusion transcript (*bottom*). N-*myc* intron and flanking sequences are shown as a line, exon noncoding sequences as open boxes, and exon-coding sequences as closed boxes. The two small arrows indicate the positions of the primers used for PCR; 5′ is to the left. (E) *Eco*RI; (B) *Bam*HI; (SII) *Sst*II; (Pv) *Pvu*II; and (Sm) *Sma*I.

using such a vector, we expected to obtain a tenfold enrichment for targeted integrations, since the frequency of obtaining neo^r colonies with the N-*myc* vector was tenfold lower than with a vector in which *neo* was transcribed from the human β-actin promoter. By screening 680 neo^r colonies with PCR, 11 targeted colonies were identified. The first two of these cell lines to be analyzed further have been confirmed as recombinants by Southern blot analysis. The overall frequency using this N-*myc* vector therefore appears to be 1 targeted colony per 60 neo^r colonies.

These experiments demonstrate that it is certainly feasible and can be relatively efficient to target mutations to specific genes of interest in ES cells. The final step of this procedure will be to transmit the mutations into the germ line via production of ES cell mouse chimeras. The phenotypic effect of specific mutations can then be studied in heterozygous and homozygous mice using standard breeding schemes and mutational analysis. In terms of the *En-2* mutation, we expect that if the cells that express *En-2* in the hind brain at midgestation require *En-2* for normal development, then homozygous *En-2* mutant embryos will die late in gestation since the hind brain is necessary for controlling a number of vital functions. However, if *En-2* is only required in the *En-2*-expressing cells of the adult cerebellum and pons region, then we expect that *En-2* homozygous mutant mice will be born that have severe defects in motor control.

OVERVIEW

Mammalian developmental genetics is clearly still in its infancy. However, the recent application of such molecular techniques as in situ analysis and PCR, combined with the exploitation of ES cells, seems sure to produce new insights into this problem. We have mainly concentrated on applying these techniques to the study of the *En* loci in part as a model system with which to test these applications and since we believe based on their expression patterns and possession of a homeo box that they play an important role in pattern formation. However, the homeo-box-containing genes are not likely to be the only genes involved in pattern formation, and we are therefore developing new approaches to identify and mutate more genes involved in this process (Gossler et al. 1989). Evidence for the role of the *En* and other genes in development should come soon from analysis of mice containing various mutations in these genes. However, for any gene we expect that mutant analysis alone will not tell the whole story, and we are therefore complementing these studies with combined cell lineage analysis in both mouse and chicken embryos. In the long run, we hope to place the *En* genes in a larger hierarchy of genes that will be found to interact and to control aspects of pattern formation.

ACKNOWLEDGMENTS

We thank the various members of our laboratories that have contributed to the work reviewed here. This work was supported by grants from the Medical Research Council (MRC), the National Cancer Institute (NCI), and the National Science and Engineering Research Council (NSERC) of Canada. A.J. is an MRC Scholar and J.R. is an NCI Research Associate. C.D. was supported by an MRC studentship, and C.M. by an NSERC Centennial studentship.

REFERENCES

Davidson, D., E. Graham, C. Sime, and R. Hill. 1988. A gene with sequence similarity to *Drosophila engrailed* is expressed during the development of the neural tube and vertebrae in the mouse. *Development* **104**: 305.

Davis, C.A. and A.L. Joyner. 1988. Expression patterns of the homeo box-containing genes *En-1* and *En-2* and the proto-oncogene *int-1* diverge during mouse development. *Genes Dev.* **2**: 1736.

Davis, C.A., S.E. Noble-Topham, J. Rossant, and A.L. Joyner. 1988. Expression of the homeo box containing gene *En-2* delineates a specific region of the developing mouse brain. *Genes Dev.* **2**: 361.

Gardner, C., D. Darnell, S. Poole, C. Ordahl, and K. Barald. 1988. Expression of an engrailed-like gene during development of the early embryonic chick nervous system. *J. Neurosci. Res.* **21**: 426.

Gay, N., S. Poole, and T. Kornberg. 1988. The *Drosophlia engrailed* protein is phosphorylated by a serine-specific protein kinase. *Nucleic Acids Res.* **16**: 6637.

Gossler, A., A.L. Joyner, J. Rossant, and W.D. Skarnes. 1989. Mouse embryonic stem cells and reporter constructs to detect developmentally regulated genes. *Science* **244**: 463.

Joyner, A.L. and G.R. Martin. 1987. *En-1* and *En-2*, two mouse genes with specific homology to the *Drosophila engrailed* gene: Expression during embryogenesis. *Genes Dev.* **1**: 29.

Joyner, A.L., W.C. Skarnes, and J. Rossant. 1989. Production of a mutation in the mouse *En-2* gene by homologous recombination in embryonic stem cells. *Nature* **338**: 153.

Joyner, A.L., T. Kornberg, K.G. Coleman, D.R. Cox, and G.R. Martin. 1985. Expression during embryogenesis of a mouse gene with sequence homology to the *Drosophila engrailed* gene. *Cell* **43**: 29.

Logan, C., H.F. Willard, J.M. Rommens, and A.L. Joyner. 1989. Chromosomal localization of the human homeo box-containing genes, *EN1* and *EN2*. *Genomics* **4**: 206.

McGinnis, W., R.L. Garber, J. Wirz, A. Kuroiwa, and W.J. Gehring. 1984. A homologous protein-coding sequence in *Drosophila* homeotic genes and its conservation in other metazoans. *Cell* **37**: 403.

Mammalian Antennapedia Class Homeo Box Genes: Organization, Expression, and Evolution

C. Kappen,[1] K. Schughart,[1] and F.H. Ruddle[1,2]

Departments of [1]Biology and [2]Human Genetics
Yale University, New Haven, Connecticut 06511

Mammalian homeo-box-containing genes were identified based on sequence similarities to homeo-box-containing genes of *Drosophila* (McGinnis et al. 1984), such as *Ultrabithorax* and *Antennapedia (Antp)*. These and other homeotic genes play crucial roles in fly development (for review, see Gehring and Hiromi 1986; Akam 1987; Ingham 1988). The function of the mammalian genes is not yet clearly understood, but accumulating evidence suggests that they contribute to the establishment of developmental patterns in the vertebrate embryo (for review, see Holland and Hogan 1988; Ruddle 1989). In both insect and mammalian homeo box genes, the homeo box region is the most highly conserved portion of the gene. It consists of 183 nucleotide bps encoding 61 amino acids. The nucleotide sequences average in similarities of approximately 65–80% between insects and mammals, and amino acid sequences in some cases are 98% similar (for review, see Scott et al. 1988). The secondary structure of the homeo domain appears to resemble a helix-turn-helix structure (Otting et al. 1988). In prokaryotes and yeast, helix-turn-helix motifs have been implicated in DNA binding by regulatory proteins (Laughon and Scott 1984; Shepherd et al. 1984). Direct experimental evidence supports a DNA-binding function for mammalian homeo domains (Fainsod et al. 1986; Odenwald et al. 1989). Studies in *Drosophila* and mouse show that the homeo domain may bind to its own regulatory region, and/or to regulatory sequences from other homeo box genes (Fainsod et al. 1986; Desplan et al. 1988; Müller et al. 1988). The interaction of homeo box genes in regulating each other was recently described in *Drosophila* (Ish-Horowicz and Pinchin 1987). A more distantly related homeo-box-containing gene, *Oct-2*, encodes a factor affecting the transcription of immunoglobulin genes (Ko et al. 1988). From these findings, it emerges that homeo box genes are controlling or coordinating the expression of effector genes important in development. The relevance of homeo box genes for the mediation of developmental events is supported by the analysis of mutations in *Drosophila* homeo box genes that dramatically affect epigenesis (Gehring and Hiromi 1986). No comparable mutations have yet been recorded for mammalian homeo box genes, but it is now possible to target these loci via homologous recombinations (Mansour et al. 1988; Joyner et al. 1989; Zimmer and Gruss 1989) in embryonic stem cells. These cells may be used as vehicles to transmit the desired mutation through the germ line (Thompson et al. 1989).

Structure of Mammalian Homeo Box Genes

The typical mammalian gene is 5–10 kb in size. In comparison, *Drosophila* genes may be eight to ten times larger. Although the homeo box proteins of insects tend to be larger, the size difference is mostly accounted for by longer *cis*-flanking domains. As a prototype for mammalian homeo box genes, the murine *Hox-2.2* gene (Schugart et al. 1988b) is depicted in Figure 1. The genomic structure of *Hox-2.2* extends over at least 10 kb, and a transcribed region of about 2.6 kb has been identified derived from two exons. The cDNA sequence contains an open reading frame encoding a predicted protein of 225 amino acids.

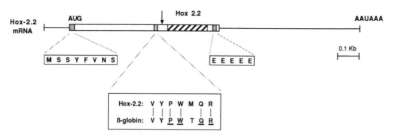

Figure 1 Structure of the *Hox-2.2* gene product. The open reading frame was predicted from cDNA sequencing (open box) (Schughart et al. 1988b); the positions of regions conserved in many homeo box proteins are indicated: the amino terminus, the hexapeptide, and the glutamic-acid-rich carboxy-terminal portion (stippled boxes); and the homeo box itself (hatched box). The *Hox-2.2* hexapeptide is compared with the corresponding sequence of β-like globins, the marked residues are thought to be involved in globin subunit interactions.

Comparisons between the amino acid sequence of *Hox-2.2* and other homeo box genes showed that conserved sequences are found in an amino-terminal octamer region, a hexapeptide 5' of the homeo domain, the homeo domain itself, and an acidic carboxy-terminal region of variable length. The hexameric region shows high-sequence similarity to a sequence present in β-type globin molecules and may be involved in protein–protein interactions. This is consistent with findings suggesting that homeo domain proteins may form functional dimers (Desplan et al. 1988). However, it is interesting to note that some homeo box proteins appear to lack the hexapeptide, for example, the murine *Hox-2.5* (Bogarad et al. 1989) and its counterpart in *Xenopus*, *XlH Box 6* (Sharpe et al. 1987).

The acidic carboxy-terminal region is composed of five glutamic acid residues in *Hox-2.2* and extends up to 15 residues in other homeo box proteins. Their function is not yet established but may involve some aspect of DNA binding and transcriptional regulation.

Chromosomal Organization of Murine and Human Homeo Box Genes

Homeo box genes bearing sequence similarity to the *Drosophila Antp* homeo box have been localized to four clusters on chromosomes 2, 6, 11, and 15 in the mouse (for review, see

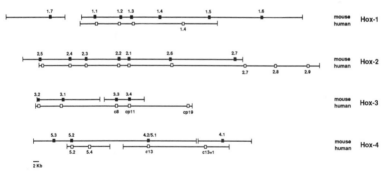

Figure 2 Structural and organization of murine and human homeobox-containing genes in clusters. The structures of the human (open boxes) and mouse (closed boxes) homeo box clusters were taken with distances inferred from Baron et al. 1987; Hart et al. 1987; Rubin et al. 1987; Breier et al. 1988; Do and Lonai 1988; Featherstone et al. 1988; Graham et al. 1988; Sharpe et al. 1988; Bogarad et al. 1989; Boncinelli et al. 1989; Duboule and Dollé 1989; Pravtcheva et al. 1989; Schughart et al. 1989; and references therein.

Schughart et al. 1988a) and on chromosomes 2, 7, 17, and 12 in humans, respectively (for review, see Ruddle et al. 1987).

A summary of the available data on the organization of homeo box genes in mice and humans is displayed in Figure 2. The longest cluster characterized to date, the *Hox-2* cluster, contains seven homeo boxes in the mouse and nine in the human, and it is likely that this is a minimum number. The genes are spaced in 5–10-kb distances and, as far as analyzed, are transcribed in the same direction. To the extent examined, expression during mouse embryonic development of all genes can be detected, and as yet no pseudogenes have been identified.

Figure 2 also shows that the order, spacing, and position of homeo box genes in mice and humans is very similar, indicating a high level of conservation during evolution. Indeed, this conservation turns out to be reflected at the level of amino acid and nucleotide sequences of the homeo box. Figure 3 contains known mouse and human amino acid sequences of Antennapedia-class homeo boxes and, for comparison, some sequences outside this class.

The homologous mouse and human sequences are more similar to each other than even the most closely related homeo boxes within either species. When the nucleotide sequences are compared, less than 10% of the nucleotide differences account for differences in amino acid sequences (Kappen et al. 1989a). This demonstrates a strong conservation of homeo box genes on the level of amino acid sequences during the divergence of the human and mouse species. The high level of sequence similarity between homologous mouse and human genes extends considerably to regions outside the homeo box (Meijlink et al. 1987; Tournier-Lasserve et al. 1989; our unpublished observations). The reasons for extensive similarity are not entirely understood, but the high sequence conservation suggests that homeo box proteins could exhibit conserved functions in both species.

Evolution of the Mammalian Homeo Box Gene Clusters

When nucleotide or amino acid sequences of murine homeo boxes are compared, it becomes apparent that the most similar boxes are located on different clusters rather than along the same cluster. They can be arranged colinearly between the clusters, so that cognate groups can be described, which include related boxes of the four clusters, such as *Hox-1.4, Hox-2.6,* and

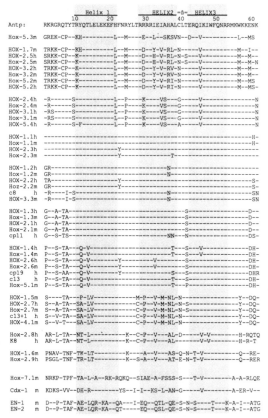

Figure 3 Comparison of amino acid sequences of murine and human homeo boxes. Amino acid sequences are displayed in single letter codes. A dash indicates residue identity to the reference sequence of the *Antp* homeo domain. Shaded areas underlie the predicted helix-forming regions. A list of the sequences with references is available upon request. Sequences of human homeo boxes were generously made available by E. Boncinelli (Naples, Italy) prior to their publication (Boncinelli et al. 1989).

Hox-4.2/5.1 toward the 3' end of the cluster (see Fig. 2). Comparisons of human homeo box sequences, respectively, would ascribe *HOX1.4*, *HOX2.6*, *c13*, and *cp19* to the same cognate group. Such cognate groups are found serially along the clusters suggesting that the four clusters originated from one ancestral cluster by duplications of large chromosomal regions (K. Schughart et al., in prep.).

On the basis of the paralogous organization of homeo boxes

between the clusters and the high sequence similarities, it is tempting to speculate that additional homeo box genes are present on the less extensively characterized clusters, *Hox-3* and *Hox-4*. Similarities among homeo box genes belonging to the same cognate group are also seen in the amino-terminal amino acid sequence, the position of the splice site, and the spacing between the homeo domain and the hexapeptide domain (Kessel et al. 1988; K. Schughart et al., in prep.). This further illustrates the paralogous relationships between homeo box clusters in both mice and humans, which extend as well to gene loci in linkage on these chromosomes. For example, collagen loci are present on chromosomes 2, 7, 17, and 12 in humans. Cytokeratins have been mapped to chromosomes 17 and 12, and desmin, another member of the intermediate filament family, is located on chromosome 2. Granulocyte colony-stimulating factor, interleukin-6, and interleukin-1 are related by sequence similarities and map to chromosomes 17, 7, and 2, respectively (for review, see Ruddle et al. 1987; Schughart et al. 1988a; Kappen et al. 1989b).

These paralogies are consistent with the occurrence of at least two chromosome region duplication events in the ancestors of mammals. Accumulating data suggest that the organization of homeo box clusters in *Xenopus* and *Zebrafish* may be very similar to the arrangement in mammals. This would place the timepoint for the presumed duplications very early in the evolution of vertebrates. The ancestral linkage group that served as a precursor for these duplication events most likely arose by an expansion process involving local gene duplications in the middle and at the extremes of the primordial cluster (Kappen et al. 1989a).

Comparisons between vertebrate and *Drosophila* homeo box genes suggest that a more distant line of evolution took place in insects. Some of the *Drosophila* homeo box genes are transcribed in opposite directions suggesting independent duplication and inversion events in arthropods.

In summary, these lines of evidence suggest that the evolution of homeo box gene clusters arose from a small gene set within an ancient linkage group that then evolved further in the arthropod and vertebrate lineages. In the lineages leading to vertebrates, more loci were added to the cluster by gene duplications, and, subsequently, the number of genes was further increased by duplication events involving entire clusters. As discussed previously (Schughart et al. 1988a; Kappen et al.

1989a,b), the increase in the number of regulatory elements might have been a crucial step in the evolution of vertebrate species.

Homeo Box Gene Expression Patterns in the Central Nervous System

The expression patterns of homeo box genes in the developing and the adult mouse have been addressed in numerous studies (for review, see Dressler and Gruss 1988; Holland and Hogan 1988) and cannot be adequately dealt with here. However, these analyses confirmed initial observations (Utset et al. 1987; for review, see Fienberg et al. 1987) that each homeo box is expressed in the central nervous system (CNS) with a specific anterior boundary, and that the positions of these anterior limits follow the order of the homeo boxes on the chromosome. For the *Hox-1* cluster of the mouse, Gaunt et al. (1988) demonstrated progressively more anterior boundaries of the expression of *Hox-1.2, -1.3, -1.4,* and *-1.5.*

Similar results were obtained by Graham et al. (1989) and Dollé and Duboule (1989) for the *Hox-2* and the *Hox-4/5* clusters. Thus, the expression patterns of homeo boxes in the CNS are correlated to the relative position on the chromosome in that the 5' extreme defines the more posteriorly and that the 3' extreme defines the more anteriorly located boundaries of anterior expression. However, it is interesting to point out that cognate homeo boxes on different clusters do not show identical expression patterns. For example, *Hox-3.1* and *Hox-2.4* belong to the same cognate group, but their anterior limits of expression vary. The functional significance of the correlation between chromosomal positions and expression patterns for embryonal development of the vertebrate embryo is not yet understood, but insight may be obtained by the experimental manipulation of homeo box gene expression using transgenic mice.

ACKNOWLEDGMENTS

We thank Mrs. Marie Siniscalchi for the preparation of the manuscript. This work was supported by grant GM-09966 from the National Institutes of Health. C.K. and K.S. were recipients of postdoctoral fellowships of the Deutsche Forschungsgemeinschaft (FRG).

REFERENCES

Akam, M. 1987. The molecular basis for metameric pattern in the *Drosophila* embryo. *Development* **101:** 1.

Baron, A., M.S. Featherstone, R.E. Hill, A. Hall, B. Galliot, and D. Duboule. 1987. *Hox-1.6:* A mouse homeo-box-containing gene member of the *Hox-1* complex. *EMBO J.* **6:** 2977.

Bogarad, L.D., M.F. Utset, A. Awgulewitsch, T. Miki, C.P. Hart, and F.H. Ruddle. 1989. The developmental expression pattern of a new murine homeo box gene: *Hox-2.5. Dev. Biol.* **133:** 537.

Boncinelli, E., D. Acampora, M. Pannese, M. D'Esposito, R. Somma, G. Gaudino, A. Stornaiuolo, M. Cafiero, A. Faiella, and A. Simeone. 1989. Organization of human class I homeobox genes. *Genome* (in press).

Breier, G., G.R. Dressler, and P. Gruss. 1988. Primary structure and developmental expression pattern of *Hox 3.1*, a member of the murine *Hox* 3 homeobox gene cluster. *EMBO J.* **7:** 1329.

Desplan, C., J. Theis, and P.H. O'Farrell. 1988. The sequence specificity of homeodomain-DNA interaction. *Cell* **54:** 1081.

Dollé, P. and D. Duboule. 1989. Two gene members of the murine *HOX-5* complex show regional and cell-type specific expression in developing limbs and gonads. *EMBO J.* **8:** 1507.

Dressler, G.R. and P. Gruss. 1988. Do multigene families regulate vertebrate development? *Trends Genet* **4:** 214.

Duboule, D. and P. Dollé. 1989. The structural and functional organization of the murine *HOX* gene family network resembles that of *Drosophila* homeotic genes. *EMBO J.* **8:** 1497.

Fainsod, A., L.D. Bogarad, T. Ruusala, M. Lubin, D.M. Crothers, and F.H. Ruddle. 1986. The homeo domain of a murine protein binds 5′ to its own homeo box. *Proc. Natl. Acad. Sci.* **83:** 9532.

Featherstone, M.S., A. Baron, S.J. Gaunt, M. Mattei, and D. Duboule. 1988. *Hox-5.1* defines the homeobox-containing gene locus on mouse chromosome 2. *Proc. Natl. Acad. Sci.* **85:** 4760.

Fienberg, A.A., M.F. Utset, L.D. Bogarad, C.P. Hart, A. Awgulewitsch, A. Ferguson-Smith, A. Fainsod, M. Rabin, and F.H. Ruddle. 1987. Homeo box genes in murine development. *Curr. Top. Dev. Biol.* **23:** 233.

Gaunt, S.J., P.T. Sharpe, and D. Duboule. 1988. Spatially restricted domains of homeo-gene transcripts in mouse embryos: Relation to a segmented body plan. *Dev. Biol.* **104:** 169.

Gehring, W.J. and Y. Hiromi. 1986. Homeotic genes and the homeobox. *Annu. Rev. Genet.* **20:** 147.

Graham, A., N. Papalopulu, and R. Krumlauf. 1989. The murine and *Drosophila* homeobox gene complexes have common features of organization and expression. *Cell* **57:** 379.

Graham, A., N. Papalopulu, J. Lorimer, J.H. McVey, E.G.D. Tuddenham, and R. Krumlauf. 1988. Characterization of a murine homeo box gene, *Hox-2.6*, related to the *Drosophila Deformed* gene. *Genes Dev.* **2:** 1424.

Hart, C.P., A. Fainsod, and F.H. Ruddle. 1987. Sequence analysis of the murine *Hox-2.2, -2.3,* and *-2.4* homeo boxes: Evolutionary and structural comparisons. *Genomics* 1: 182.

Holland, P.W.H. and B.L.M. Hogan. 1988. Expression of homeo box genes during mouse development: A reiew. *Genes Dev.* 2: 773.

Ingham, P.W. 1988. The molecular genetics of embryonic pattern formation in *Drosophila. Nature* 335: 25.

Ish-Horowicz, D. and S.M. Pinchin. 1987. Pattern abnormalities induced by ectopic expression of the *Drosophila* gene *hairy* are associated with repression of *ftz* transcription. *Cell* 51: 405.

Joyner, A.L., W.C. Skarnes, and J. Rossant. 1989. Production of a mutation in mouse *En-2* gene by homologous recombination in embryonic stem cells. *Nature* 338: 153.

Kappen, C., K. Schughart, and F.H. Ruddle. 1989a. Two steps in the evolution of vertebrate Antennapedia class homeobox genes. *Proc. Natl. Acad. Sci.* (in press).

———. 1989b. Organization and expression of homeobox genes in mouse and man. *Ann. N.Y. Acad. Sci.* (in press).

Kessel, M., M. Fibi, and P. Gruss. 1988. Organization of homeodomain proteins. In *Cellular factors in development and differentiation: Embryos, teratocarcinomas, and differentiated tissues,* p. 93. A.R. Liss, New York.

Ko, H.S., P. Fast, W. McBride, and L.M. Staudt. 1988. A human protein specific for the immunoglobulin octamer DNA motif contains a functional homeobox domain. *Cell* 55: 135.

Laughon, A. and M.P. Scott. 1984. Sequence of a *Drosophila* segmentation gene: Protein structure homology with DNA binding proteins. *Nature* 310: 25.

Mansour, S.L., K.R. Thomas, and M.R. Capecchi. 1988. Disruption of the proto-oncogene *int-2* in mouse embryo-derived stem cells: A general strategy for targeting mutations to non-selectable genes. *Nature* 336: 348.

McGinnis, W., R.L. Garber, J. Wirz, A. Kuroiwa, and W.J. Gehring. 1984. A homologous protein-coding sequence in *Drosophila* homeotic genes and its conservation in other metazoans. *Cell* 37: 403.

Meijlink, F., R. deLaaf, P. Verrijzer, O. Destrée, V. Kroezen, J. Hilkens, and J. Deschamps. 1987. A mouse homeobox containing gene on chromosome 11: Sequence and tissue-specific expression. *Nucleic Acids Res.* 15: 6773.

Müller, M., M. Affolter, W. Leupin, G. Otting, K. Wüthrich, and W.J. Gehring. 1988. Isolation and sequence-specific DNA binding of the *Antennapedia* homeodomain. *EMBO J.* 7: 4299.

Odenwald, W.F., J. Garbern, H. Arnheiter, E. Tournier-Lasserve, and R.A. Lazzarini. 1989. The *Hox-1.3* homeo box protein is a sequence specific DNA-binding phosphoprotein. *Genes Dev.* 3: 158.

Otting, G., Y.-Q. Qian, M. Müller, M. Affolter, W. Gehring, and K. Wüthrich. 1988. Secondary structure determination for the *Antennapedia* homeodomain by nuclear magnetic resonance and evidence for a helix-turn-helix motif. *EMBO J.* 7: 4305.

Pravtcheva, D., M. Newman, L. Hunihan, P. Lonai, and F.H. Ruddle. 1989. Chromosome assignment of the murine *Hox-4.1* gene. *Genomics* (in press).

Rubin, M.R., W. King, L.E. Toth, I.S. Sawczuk, M.S. Levine, P.D. D'Eustachio, and M.C. Nguyen-Huu. 1987. Murine *Hox-1.7* homeobox gene: Cloning, chromosomal location, and expression. *Mol. Cell. Biol.* **7:** 3836.

Ruddle, F.H. 1989. Genomics and evolution of murine homeo box genes. In *The physiology of growth* (ed. J.M. Tanner and M.A. Priest). Cambridge University Press, England. (In press.)

Ruddle, F.H., C.P. Hart, M. Rabin, A.C. Ferguson-Smith, and D. Pravtcheva. 1987. Comparative genetic analysis of homeo-box genes in mouse and man. In *Human genetics* (ed. F. Vogel and K. Sperling), p. 419. Springer-Verlag, Heidelberg.

Schughart, K., C. Kappen, and F.H. Ruddle. 1988a. Mammalian homeobox-containing genes: Genome organization, structure, expression and evolution. *Br. J. Cancer* **58:** 9.

Schughart, K., M.F. Utset, A. Awgulewitsch, and F.H. Ruddle. 1988b. Structure and expression of *Hox-2.2*, a murine homeobox-containing gene. *Proc. Natl. Acad. Sci.* **85:** 5582.

Schughart, K., D. Pravtcheva, M.S. Newman, L.W. Hunihan, Z. Jiang, and F.H. Ruddle. 1989. Isolation and regional localization of the murine homeobox-containing gene *Hox-3.3* to mouse chromosome region 15E. *Genomics* (in press).

Scott, N.P., J.W. Tamkun, and G.W. Hartzell. 1988. The structure and function of of the homeodomain. *Biochim. Biophys. Acta Rev. Cancer* (in press).

Sharpe, C.R., A. Fritz, E.M. DeRobertis, and J.B. Gurdon. 1987. A homeobox-containing marker of posterior neural differentiation shows the importance of predetermination in neural induction. *Cell* **50:** 749.

Sharpe, P.T., J.R. Miller, E.P. Evans, M.D. Burtenshaw, and S.J. Gaunt. 1988. Isolation and expression of a new mouse homeobox gene. *Development* **102:** 397.

Shepherd, J.C., W. McGinnis, A.E. Carrasco, E.M. DeRobertis, and W.J. Gehring. 1984. Fly and frog homeo domains show homologies with yeast mating type regulatory proteins. *Nature* **310:** 70.

Thompson, S., A.R. Clarke, A.M. Pow, M.L. Hooper, and D.W. Melton. 1989. Germ line transmission and expression of a corrected HPRT gene produced by gene targeting in embryonic stem cells. *Cell* **56:** 313.

Tournier-Lasserve, E., W.F. Odenwald, J. Garbern, J. Trojanowski, and R.A. Lazzarini. 1989. Remarkable intron and exon sequence conservation in human and mouse homeobox *Hox 1.3* genes. *Mol. Cell. Biol.* **9:** 2273.

Utset, M.F., A. Awgulewitsch, F.H. Ruddle, and W. McGinnis. 1987. Region-specific expression of two mouse homeo box genes. *Science* **235:** 1379.

Zimmer, A. and P. Gruss. 1989. Production of chimaeric mice containing embryonic stem (ES) cells carrying a homeobox *Hox-1.1* allele mutated by homologous recombination. *Nature* **338:** 150.

The Organization of the Murine *Hox* Gene Family Resembles That of *Drosophila* Homeotic Genes

B. Galliot,[1] P. Dollé,[1] and D. Duboule[2]

[1]Laboratoire de Génétique Moléculaire des Eucaryotes du CNRS
Unité 184 de Biologie et de Génie Génétique de l'INSERM
Faculté de Médecine, 67085 Strasbourg Cédex, France

[2]European Molecular Biology Laboratory, D-6900 Heidelberg
Federal Republic of Germany

Antennapedia (*Antp*)-like homeo-box-containing genes have been isolated from many different invertebrate and vertebrate species (see McGinnis 1985; Holland and Hogan 1986; Gehring 1987). In *Drosophila*, the homeo-box-containing homeotic genes are essentially dispersed in two clusters: the Bithorax complex (BX-C; Lewis 1978) and Antennapedia complex (ANT-C; Kaufmann et al. 1980), whose internal structural organizations have been shown to reflect the order in which these genes are expressed along the anterior-posterior body axis during development. Thus, the relative position of each of these genes within the complexes correlates to their expression domains along the anterior-posterior axis, and, as a result, the structures that are specified by these different domains (for review and additional references, see Harding et al. 1985; Akam 1987; Scott and Carrol 1987). In the house mouse, more than 20-*Antp*-like homeo-box-containing genes have been reported so far to lie on four major complexes: HOX-1, -2, -3 and -5, located on chromosomes 6, 11, 15, and 2, respectively. On the basis of protein sequence similarity, these genes can be grouped into subfamilies that are represented once in each cluster. Because each subfamily member is found in the same relative position within each cluster (Hart et al. 1987; Graham et al. 1988; Duboule et al. 1989), the HOX-1, -2, -3 and -5 complexes are likely to be the result of large-scale duplication events during evolution.

Accumulating evidence indicates that murine *Hox* genes (homeo genes) play a crucial role in embryonic and fetal development. Because they are expressed during ontogeny in the same types of structures (e.g., central and peripheric nervous system, somitic and nonsomitic mesoderm derivatives, and limb buds) but in different overlapping anterior-posterior domains (for additional references, see Gaunt et al. 1988; Holland and Hogan 1988), it was suggested that vertebrate homeo proteins might serve as potential cues along the rostro-caudal axis of the developing animal. It was recently further proposed that, as in *Drosophila*, the ordering of the *Hox* genes along the various complexes may also reflect the antero-posterior distribution of their expression domains and that gene members of the same subfamily might therefore display coincident anterior expression boundaries (Gaunt et al. 1988). On the basis of these observations and on significant similarities to sequences and relative domains of expression of homeotic genes of *Drosophila*, we suggest that both the structural and functional organization of the homeo-box-containing gene families have been conserved between insects and vertebrates. In this communication, we would like to use the mouse *Hox-1.4* gene to illustrate some aspects of this proposal.

Hox Clusters and Subfamilies

Because of the evolutionary relationships between the four murine *Hox* clusters, there will be a maximum of four homeo genes derived from the same ancestor gene. These genes, located in a vertical box when the complexes are aligned (see Fig. 2) are therefore expected to show a high degree of similarity. This is illustrated in Figure 1, where the mouse *Hox-1.4* open reading frame (B. Galliot et al., in prep.) is aligned with the *Hox-2.6* (Graham et al. 1989) and *Hox-5.1* (Featherstone et al. 1988) genes. These three genes are members of the same subfamily (see Fig. 2), as shown by the extension of the similarities outside the homeo domain itself.

Structural Conservation between *Drosophila* and Mouse Genes

Some of the *Drosophila* homeotic genes can be considered as homologs of a murine subfamily based on sequence similarity. Figure 1 shows that the *Drosophila* homeotic gene *Deformed (Dfd)* (Regulski et al. 1987) also possesses the blocks of high similarity that are present in the members of the *Hox-1.4* subfamily. Similarly, although to a lesser extent, the labial gene can be considered as a member of the *Hox-1.6*-like family (Baron et al. 1987; Duboule et al. 1989), and such homologies

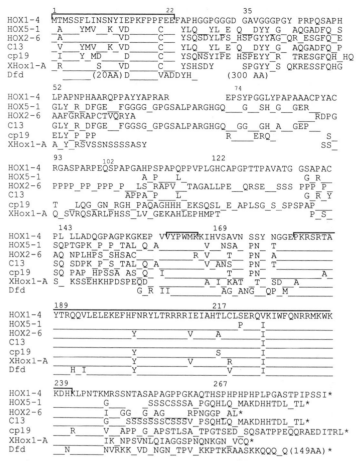

Figure 1 Comparison of the *Hox-1.4* predicted protein with other related homeo gene products. The murine *Hox-1.4* sequence is used as the basis for comparison with other sequences from mouse (*Hox-5.1* and *Hox-2.6*) human (C13, cp19), *Xenopus* (XHox1A) and *Drosophila* (*Dfd*) (B. Galliot et al., in prep.). Hyphens represent identical amino acids whereas spaces are used to facilitate the best possible alignment. The three main regions of consensus are underlined: the amino-terminal part, the hexapeptide, and the homeo domain.

with *Hox* genes can be extended to most of the *Drosophila* homeotic gene members of either BX-C or ANT-C (Fig. 2). It then becomes obvious that the relationship between the structure of a gene and its relative position within its cluster is identical in both *Drosophila* and vertebrates, strongly suggesting a common origin for the murine HOX complexes and the *Drosophila* BX-C and ANT-C (Duboule and Dollé 1989; Graham et al. 1989).

A Molecular Representation of the Vertebrate Body Axis

Hox-1.4 is expressed during fetal development in various structures of ectodermal and mesodermal origin up to an anterior

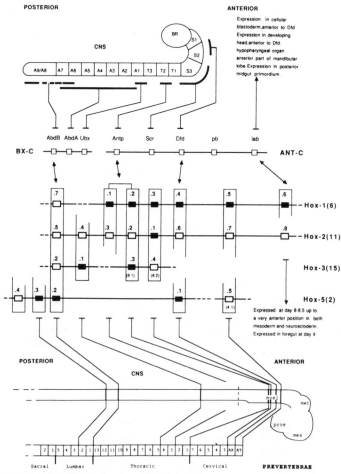

Figure 2 (*See facing page for legend.*)

limit best visualized in the prevertebral column ([pv2] Gaunt et al. 1988; B. Galliot et al., in prep.). This boundary is located posteriorly to that of the *Hox-1.5* gene (pv1) but anteriorly to that of the *Hox-1.3* gene. Therefore, the anterior-expression boundary of a given *Hox* gene is determined by its relative position within its complex (Gaunt et al. 1988), according to the rule: 3′ → anterior and 5′ → posterior. Consequently, genes belonging to the same subfamily may be expected to have comparable respective expression domains. Thus, the vertebrate genome contains a direct molecular representation of the major body axis. This is highly comparable with the situation that exists in *Drosophila* (see, e.g., Fig. 2 and Harding et al. 1985) and suggests that the developmental control mechanism achieved, both in flies and vertebrates, by the homeo-box-containing gene products might have been conserved. The similarities found at the level of the homeo boxes probably reflect the conservation of target-binding sites required in a regulatory network (see Desplan et al. 1988; Hoey and Levine 1988; and references therein). In this case, functional mechanisms may have been conserved although their specific roles may be different. The definition in vertebrates of genes regulating or under the regulation of the *Hox* network will elucidate these questions.

Figure 2 Schematic representation of the possible correlation between the *Drosophila* homeotic gene complexes and the murine (vertebrate) *Hox* gene network. (*Top*) Domains of expression of *Drosophila* homeotic genes, members of either BX-C or ANT-C, in the embryonic central nervous system (CNS). Other nonhomeotic genes (*zen, bic,* and *ftz*) located within the ANT-C are not indicated for clarity. (*Middle*) HOX complexes with genes (closed boxes) that have been studied by comparative in situ hybridization experiments and whose expression domains (or, at least, the position of their anterior boundaries) have been defined. These boundaries are probably representative of those of all the genes belonging to the same subfamily (vertical boxes). Genes within the same rectangle are expected to share comparable antero-posterior expression boundaries or, at least, comparable antero-posterior relative expression boundaries within their respective complexes (see Gaunt et al. 1988). (*Bottom*) Antero-posterior boundaries of expression of these genes along the fetal central nervous system and prevertebral column. In both structures, a unique boundary is given for each gene subfamily without considering slight variations that might occur within a particular subfamily (see text). *Top* is linked to the *Middle* and *Bottom* parts with arrows indicating significant amino-acid similarities between *Drosophila* and murine genes. See Duboule and Dollé (1989) and references therein. (Reprinted, with permission, from Duboule and Dollé 1989.)

81

ACKNOWLEDGMENTS
We thank Pierre Chambon for his support and Hilary Davies for preparing the manuscript. This work was carried out with grants from the CNRS, INSERM, ARC, and the Foundation pour la Recherche Médicale.

REFERENCES

Akam, M. 1987. The molecular basis for metameric pattern in the *Drosophila* embryo. *Development* **101**: 1.

Baron, A., M.S. Featherstone, R.E. Hill, A. Hall, B. Galliot, and D. Duboule. 1987. *Hox-1.6*: A mouse homeo-box-containing gene member of the *Hox-1* complex. *EMBO J.* **6**: 2977.

Desplan, C., J. Theis, and P.H. O'Farrell. 1988. The sequence specificity of homeodomain-DNA interaction. *Cell* **54**: 1081.

Duboule, D. and P. Dollé. 1989. The structural and functional organization of the murine *Hox* gene family resembles that of *Drosophila* homeotic genes. *EMBO J.* **8**: 1497.

Duboule, D., B. Galliot, A. Baron, and M.S. Featherstone. 1989. Murine homeo-genes: Some aspects of their organization and structure. *NATO ASI Ser.* Ser. H Cell Biol. **26**: 97.

Featherstone, M.S., A. Baron, S.J. Gaunt, M.G. Mattei, and D. Duboule. 1988. *Hox-5.1* defines a homeo-box containing gene locus on mouse chromosome 2. *Proc. Natl. Acad. Sci.* **85**: 4760.

Gaunt, S.J., P.T. Sharpe, and D. Duboule. 1988. Spatially restricted domains of homeo-box transcripts in mouse embryos: Relation to a segmented body plan. *Development* (suppl.) **104**: 169.

Gehring, W.J. 1987. Homeo boxes in the study of development. *Science* **236**: 1245.

Graham, A., N. Papalopulu, and R. Krumlauf. 1989. The murine and *Drosophila* homeobox gene complexes have common features of organization and expression. *Cell* **57**: 367.

Graham, A., N. Papalopulu, J. Lorimer, J.-H. McVey, E.G.D. Tuddenham, and R. Krumlauf. 1988. Characterization of a murine homeo box gene, *Hox-2.6*, related to the *Drosophila Deformed* gene. *Genes Dev.* **2**: 1424.

Harding, K., C. Weddeen, W. McGinnis, and M. Levine. 1985. Spatially regulated expression of homeotic genes in a *Drosophila*. *Science* **229**: 1236.

Hart, C.P., A. Fainsod, and F.H. Ruddle. 1987. Sequence analysis of the murine *Hox-2.2, -2.3,* and *-2.4* homeo-boxes: Evolutionary and structural comparisons. *Genomics* **1**: 182.

Hoey, T. and M. Levine. 1988. Divergent homeobox proteins recognize similar DNA sequences in *Drosophila*. *Nature* **332**: 858.

Holland, P.W.H. and B.L.M. Hogan. 1986. Phylogenetic distribution of *Antennapedia*-like homeo boxes. *Nature* **321**: 251.

———. 1988. Expression of homeo box genes during mouse development: A review. *Genes Dev.* **2**: 773.

Kaufman, T.C., R. Lewis, and B.T. Wakimoto. 1980. Cytogenetic analysis of chromosome 3 in *Drosophila melanogaster* the homeotic gene complex in polytene chromosome interval 84A,B. *Genetics* **94**: 115.

Lewis, E.B. 1978. A gene complex controlling segmentation in *Drosophila. Nature* **276**: 565.

McGinnis, W. 1985. Homeo box sequences of the Antennapedia class are conserved only in higher animal genomes. *Cold Spring Harbor Symp. Quant. Biol.* **50**: 263.

Regulski, M., N. McGinnis, R. Chadwick, and W. McGinnis. 1987. Developmental and molecular analysis of *Deformed*; a homeotic gene controlling *Drosophila* head development. *EMBO J.* **6**: 767.

Scott, M.P. and S.B. Carroll. 1987. The segmentation of homeotic gene network in early *Drosophila* development. *Cell* **51**: 689.

The Murine and *Drosophila* Homeo Box Clusters Are Derived from a Common Ancestor Based on Similarities in Structure and Expression

A. Graham,[1] N. Papalopulu,[1] P. Hunt,[1]
M.H. Sham,[1] L. Simmoneau,[1] M. Cook,[1]
A. Bradley,[2] and R. Krumlauf[1]

[1]Laboratory of Eukaryotic Molecular Genetics, National Institute for Medical Research, London NW7 1AA, United Kingdom
[2]Institute for Molecular Genetics, Baylor College of Medicine Houston, Texas 77030

Genes from *Drosophila* implicated in specifying segmental pattern and segment identity have been found to contain a common element, termed the homeo box (for review, see Gehring 1987). This element has been found in the genomes of many nonsegmented species and also in vertebrates (McGinnis 1985; Holland and Hogan 1986), suggesting that it does not simply correlate with a segmented body plan, but that it may also have a broader role in development. The homeo box has also been found in *Drosophila* genes exhibiting patterns of expression that suggest roles in positional information independent of the process of segmentation. Thus, it would appear that homeo-box-containing genes are involved not only in the process of segmentation, but also in specifying positional information. Patterns of expression of mouse homeo box genes display similar but nonidentical overlapping domains, consistent with a role in interpretation of positional information along the anterior-posterior axis. Therefore, it seems possible that the presence of the homeo domain in vertebrates is not merely the recruitment of a protein motif encoding DNA-binding domains, but also the conservation of a system for specifying positional information along the anterior-posterior axis of the embryo. Our analysis of the mouse *Hox-2* cluster supports this idea.

RESULTS

Organization and Expression of the Hox-2 Complex along the Anterior-Posterior Axis

The *Hox-2* homeo box cluster was initially isolated by Hart et al. (1985) and mapped to chromosome 11. We have recently reported the isolation of cosmid clones spanning the Hox-2 complex, which contained seven homeo box genes (Graham et al. 1988). Using probes for each of these genes, we have investigated the patterns of expression in 12.5-day-old mouse embryos by in situ hybridization. Complex domains of expression are observed in both ectodermal and mesodermal derivatives; however, we have observed overlapping but distinct domains of expression for each gene in the central nervous system (CNS). In these domains, we see no clear posterior boundary of expression, and we have focused on the different anterior boundaries of expression for each member of the Hox-2 complex. All seven genes have been directly compared in serial sections, using relative distances from morphological structures to map the

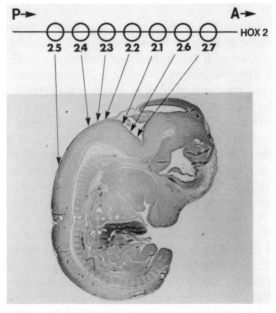

Figure 1 Diagrammatical representation of the anterior boundaries of expression in the CNS for the Hox-2 complex and their correlation with gene position in the cluster. (Reprinted, with permission, from Graham et al. 1989).

boundaries. Starting with *Hox-2.5*, we find that as one progresses through the Hox-2 complex in a 5′ to 3′ direction, the limit of expression of each gene maps to a progressively more anterior region of the CNS (Graham et al. 1989). Many of the anterior limits of expression map within the hindbrain region, and a summary of the in situ results is illustrated in Figure 1. On the basis of these findings, we conclude that the position of a gene in the *Hox-2* cluster reflects its relative domain of expression along the anterior-posterior axis of the embryo in the CNS. Similar correlations between position and expression of homeo box genes from other mouse clusters have also been described previously (Gaunt et al. 1988; Duboule and Dollé 1989). This correlation in expression is analogous to the *Drosophila* Bithorax complex (BX-C) and the Antennepedia complex (ANT-C) homeo box gene clusters, where there is a correlation between the relative position of a gene within a cluster, its domain of expression, and its effects along the anterior-posterior axis of the embryo (for review, see Akam 1987).

Comparison between Murine and *Drosophila* Homeo Box Clusters

This similarity between mouse and *Drosophila* homeo box complexes is not limited only to patterns of expression. The *Hox-2.6* gene shares substantial sequence identity in multiple regions of the protein not only with certain mouse (Featherstone et al. 1988) and vertebrate homeo box genes, but also with the *Drosophila Deformed (Dfd)* gene (Regulski et al. 1987; Graham et al. 1988). On the basis of these identities and the similarities in expression patterns, we examined how other genes in the mouse *Hox* clusters might be related to other *Drosophila* homeo box genes. Initially, we had identified seven members of the Hox-2 complex; however, in cosmid walking and yeast artificial chromosome libraries (in collaboration with H. Lehrach), we have identified two more members of the cluster termed *Hox-2.8* and *Hox-2.9*. Comparison of the *Hox-2* homeo domain sequences reveals that there are significant identities with many other mouse homeo domains. This allows one to separate the known sequenced murine genes into subfamilies of related sequences. These subfamilies are defined by common amino acid changes and the position of the changes when compared with *Antennapedia (Antp)*. When all of the related mouse subfamilies are aligned as shown in Figure 2, it is clear that each of the related genes from different mouse homeo box clusters

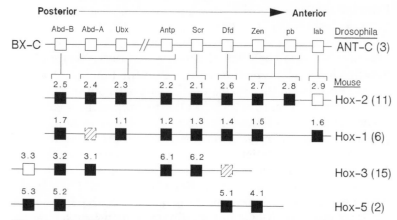

Figure 2 Alignment representing relationships between the *Drosophila* and murine homeo box gene clusters. Members of a subfamily are represented by vertical rows of boxes. (Closed boxes) Sequenced domains; (open boxes) identified but not completely sequenced; (hatched boxes) predicted genes based on other clusters or human organization. Numbers in parenthesis represent chromosomal localization, and numbers above the boxes are those assigned as genes were originally isolated.

has a similar position and order in its respective cluster. The alignment also reveals that not all genes or subfamilies are represented in each mouse cluster. This type of analysis supports the idea from several groups that the four major mouse clusters arose by duplication and divergence from a common single cluster.

If we extend this comparison to include the *Drosophila* homeo domains, it is clear that many of the sequences show similarity to different specific subfamilies. When we align the *Drosophila* genes with their most closely related mouse genes, the physical order of the *Drosophila* homeo box genes along the chromosome is identical with that of the related mouse gene (Duboule and Dollé 1989; Graham et al. 1989), as illustrated at the top of Figure 2. These results suggest that the murine and *Drosophila* homeo box gene clusters are derived from a common ancestor that arose before the divergence of lineages, which gave rise to arthropods and vertebrates. This apparent common evolutionary origin and the correlations in domains of expression along the anterior-posterior axis further supports the idea that the vertebrate genes may be playing a role in specifying positional information.

Functional Analysis of *Hox-2* Genes in Vertebrate Embryos

To approach more directly the functional role of the *Hox-2* genes during development, we have been using a variety of approaches. Embryonic stem cells have been transfected with constructs to disrupt the endogenous *Hox-2.6* gene by homologous recombination. Candidate clones from these experiments are being injected into blastocytes in attempts to generate germline chimeric mice carrying the disrupted allele. These animals can then be used to generate lines for a null phenotype. Altered patterns of expression in transgenic mice are also being used to generate dominant phenotypes. We have found that *Hox-2.1* gene is regulated in part at the posttranscriptional level in teratocarcinoma and embryonic stem cells. This control is partially mediated by sequences in the 3′ untranslated region of the RNA that render the RNA unstable. Removal of this *cis* sequence results in increased stability and accumulation of the RNA. We have used constructs expressing the more stable RNA from its own promoter to examine the effects of overexpression of the *Hox-2.1* gene during development.

Finally, on the basis of the high degree of conservation in both proteins and domains of expression shared between mouse, *Xenopus,* and chick, we have been injecting RNAs encoding mouse homeo domain proteins into developing *Xenopus* embryos. Three *Hox-2* RNAs, *Hox-2.1, Hox-2.6,* and *Hox-2.7,* generate a series of altered phenotypes. One of these phenotypes results in a bifurcation of the neural tube, generating a duplication of the posterior structures. This does not appear to be a transformation of cell types but a duplication of normal structures such as somites, the notocord, and the neural tube. Embryos appear to be affected in the early stages of gastrulation; however, because head structures appear normal, this is most likely not a problem in the initiation of gastrulation. Similar phenotypes have been observed in embryos that do not invaginate properly during gastrulation. We are currently examining the complex phenotypes from these injections to determine to what extent they correlate with different homeo box genes in the *Hox-2* cluster.

ACKNOWLEDGMENTS

We thank Denis Duboule for discussions of his work before publication. N.P. was supported by the Greek State Scholarship Foundation. P.H. and A.G. are supported by M.R.C. post-graduate studentships.

REFERENCES

Akam, M.E. 1987. The molecular basis for metameric pattern in the *Drosophila* embryo. *Development* **101**: 1

Duboule, D. and P. Dollé. 1989. The structural and functional organization of the murine Hox gene family resembles that of *Drosophila* homeotic genes. *EMBO J.* **8**: 1497.

Featherstone, M.S., A. Baron, S.J. Gaunt, M.G. Mattei, and D. Duboule. 1988. *Hox-5.1* defines a homeobox-containing locus on mouse chromosome 2. *Proc. Natl. Acad. Sci.* **85**: 4760.

Gaunt, S.J., P. Sharpe, and D. Duboule. 1988. Spatially restricted domains of homeogene transcripts in mouse embryos: Relation to a segmented body plan. *Development* (suppl.) **104**: 169.

Gehring, W.J. 1987. Homeo boxes in the study of development. *Science* **236**: 1245.

Graham, A., N. Papalopulu, and R. Krumlauf. 1989. The murine and *Drosophila* homeobox gene complexes have common features of organisation and expression. *Cell* **57**: 367.

Graham, A., N. Papalopulu, J. Lorimer, J. McVey, E.G.D. Tuddenham, and R. Krumlauf. 1988. Characterization of a murine homeo box gene, *Hox-2.6*, related to the *Drosophila Deformed* gene. *Genes Dev.* **2**: 1424.

Hart, C.P., A. Awgulewitsch, A. Fainsod, W. McGinnis, and F.H. Ruddle. 1985. Homeobox gene complex on mouse chromosome 11: Molecular cloning, expression in embryogenesis and homology to a human homeobox locus. *Cell* **43**: 9.

Holland, P. and B.M.L. Hogan. 1986. Phylogenetic distribution of *Antennapedia*-like homeoboxes. *Nature* **321**: 251.

McGinnis, W. 1985. Homeo box sequences of the Antennapedia class are conserved only in higher animals. *Cold Spring Harbor Symp. Quant. Biol.* **50**: 263.

Regulski, M., N. McGinnis, R. Chadwick, and W. McGinnis. 1987. Developmental and molecular analysis of *Deformed;* a homeotic gene controlling *Drosophila* head development. *EMBO J.* **6**: 767.

The *Hox-1.1* Promoter Directs Expression to a Specific Region of the Embryo in Transgenic Mice

A.W. Püschel, R. Balling, G.R. Dressler, and P. Gruss

Department of Molecular Cell Biology, Max-Planck-Institute of Biophysical Chemistry, 3400 Göttingen, Federal Republic of Germany

The murine homeo-box-containing genes (*Hox* genes) are expressed during embryogenesis in specific overlapping regions along the anterior-posterior axis (Dressler and Gruss 1988; Holland and Hogan 1988). Because many homeo domain proteins bind specific DNA sequences (Levine and Hoey 1988), it is generally assumed that genes such as *Drosophila Antennapedia, fushi tarazu, engrailed,* and *even-skipped* encode transcription factors regulating developmental processes. The first types of homeo domain proteins with known functions are the well-characterized mammalian transcription factors Oct-1 and Oct-2 (Herr et al. 1988). Although highly specific *Hox* gene expression during embryogenesis is indicative for participation in developmental processes, it remains to be proven whether they have a similar function. The *Hox-1.1* gene, for example, is first expressed at the time of primitive streak retraction when somites are formed from the paraxial mesoderm (Fig. 1b) (Mahon et al. 1988). During the subsequent stages of development, *Hox-1.1* expression selectively persists in the neural tube, spinal ganglia, and sclerotomes. Identification of the *cis*-acting sequences regulating this spatiotemporal expression profile will address the molecular mechanisms responsible for pattern formation during murine embryogenesis.

Expression in Transgenic Mice
Expression from *Hox-1.1* promoter sequences was analyzed in transgenic mice using 3.6 kb of potential promoter sequences linked to the *Escherichia coli lacZ* gene and 1.7 kb of *Hox-1.1* 3' untranslated sequences, including the poly(A) addition site (Fig. 1a). This construct, *m6lacZ1*, was injected into fertilized

oocytes, and transgenic mice were generated. *m6lacZ1* directed expression of β-galactosidase to a specific region of the embryo. Similar to the endogenous *Hox-1.1*, *m6lacZ1* expression is first detectable around day 7.5 postcoitum in the allantois. Subsequently, the transgene is turned on first in the ectoderm (Fig. 1b,c) and then in the mesoderm (Fig. 1d). At this early stage of

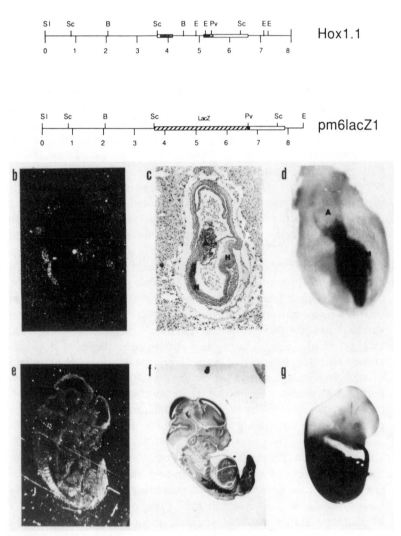

Figure 1 (*See facing page for legend.*)

development (day 7.5–8.5), the transgene is regulated in parallel to the endogenous *Hox-1.1* gene. At day 12.5 postcoitum, the time of maximal *Hox-1.1* expression during embryogenesis, the *m6lacZ1* transgene is strongly expressed in spinal cord and dorsal root ganglia, starting at the level of the C5 ganglion and extending to the posterior end of the embryo (Fig. 1f,g). Expression in the mesoderm is first seen approximately six segments caudally at the level of the fourth thoracic prevertebrae and continues caudally in almost all cell types. Although there is an anterior border similar to the *Hox-1.1* gene, there is no posterior restriction of expression. In contrast, *Hox-1.1* mRNA levels decrease caudal to the 13th thoracic vertebrae to levels undetectable by in situ hybridization (Mahon et al. 1988). Transgene activity is, however, uniformly high in the tail. Thus, the *m6lacZ1* transgene is not regulated appropriately, and no posterior limit of expression is specified. The pattern observed for *m6lacZ1* at day 12.5 is reminiscent of the early expression of *Hox-1.1* between days 8 and 9 with high expression in the posterior part of the embryo. Around day 9, expression patterns of *Hox-1.1* and *m6lacZ1* diverge. Whereas *Hox-1.1* ex-

Figure 1 Expression of *m6lacZ1* in transgenic mice. (*a*) Genomic organization of *Hox-1.1* (*upper panel*) and construct used to generate transgenic mice (*lower panel*). Exons are indicated by bars: (Open bar) Nontranslated sequences; (stippled bar) coding sequences; (closed bar) homeo box; (hatched bar) *lacZ* coding sequence. Sequences between the *SacI* and *PvuII* site of *Hox-1.1* were replaced by *lacZ* coding sequences. *m6lacZ1* contains 3.6-kb 5′ and 1.7-kb 3′ *Hox-1.1* sequences. (S1) *SalI*; (Sc) *SacI*; (B) *BamHI*; (E) *EcoRI*; (Pu) *PvuII*; and numbers indicate kbp of DNA. (Dark-field image of an in situ hybridization: A sagital section of a day 8 embryo was hybridized with a *Hox-1.1*-specific ^{35}S-labeled RNA probe. Hybridization is visible in neural plate. (*c*) β-Galactosidase activity in neural plate and allantois of a day 8 transgenic embryo. Staining is most pronounced in the posterior part of the neuroectoderm. Embryos were stained with 5-bromo-4-chloro-3-indolyl-β-D galactoside, embedded in paraffin, and sectioned. (A) Allantois; (N) neuroectoderm, and (H) head fold. (*d*) β-Galactosidase activity in day 8.5 transgenic embryo (view of posterior ventral side). Staining is seen in closing neural plate and mesoderm. (A) Allantois and (M) mesoderm. (*e*) Dark-field image of an in situ hybridization: A sagital section of a day 12 embryo was hybridized with a *Hox-1.1*-specific ^{35}S-labeled RNA probe. Hybridization is detected in prevertebrae. (*f*) Parasagital section of a day 13 embryo: Whole embryos were stained, paraffin-embedded, and sectioned. Staining is detectable in prevertebrae (→). (*g*) β-Galactosidase activity in day 12 embryo: Whole embryos were dissected and stained. Activity is seen in posterior part of the embryo.

pression persists only in some cells, *m6lacZ1* activity remains high in all cells originating within the region of activity initially defined. *Hox-1.1* expression is regulated in at least two phases. Initially, *Hox-1.1* expression is established in a specific region of the embryo. Subsequently, expression is maintained in only a subset of structures (Mahon et al. 1988). This second regulatory phase is not executed for *m6lacZ1*, whereas the early phase is correctly regulated. Identification of the sequences controlling late posterior patterning and resulting in correct expression in all respects will show whether indeed these two phases of regulation exist and what molecular mechanisms are involved. Preliminary experiments show that the inclusion of additional genomic sequences is necessary to obtain correct expression of a transgene that has both an anterior and a posterior boundary of expression and is restricted to a subset of mesodermal tissues.

CONCLUSIONS

Dissecting the *cis*-acting elements responsible for the complex spatiotemporal expression pattern of *Hox-1.1* is a first step in resolving the molecular mechanisms of pattern formation in mammals. The *m6lacZ1* transgene promoter is the first promoter that is responsive to positional information rather than being tissue specific and which, although broadened in its domain of activity, reflects the regulation of the corresponding gene. A previous report has shown restricted activity of the *Hox-1.3* promoter (Zakany et al. 1988). However, the reported pattern does not correlate either in time or in space with the identified pattern of *Hox-1.3* RNA. In preliminary experiments, the addition of genomic sequences modified the pattern of *m6lacZ1* as described above to fit the distribution of *Hox-1.1* RNA. Therefore, different *cis*-acting elements are responsible for different aspects of position-dependent expression. Mapping of the position-responsive elements will facilitate the purification of factors signaling positional values. The corresponding transgenic lines will provide useful markers to assay effects of experimental manipulations or mutations during development.

REFERENCES

Dressler, G.R. and P. Gruss. 1988. Do multigene families regulate vertebrate development? *Trends Genet.* **4**: 214.

Herr, W., R.A. Sturm, R.G. Clerc, L.M. Corcoran, D. Baltimore, P.A. Sharp, M.A. Ingraham, M.G. Rosenfeld, M. Finney, G. Ruukun, and H.R. Horowitz. 1988. The POU domain: A large conserved region in the mammalian *pit-1*, *oct*-1, *oct*-2, and *Caernorhabditis elegans unc*-86 gene products. *Genes Dev.* **2:** 1513.

Holland, P.W.H. and B.L.M. Hogan. 1988. Expression of homeo box genes during mouse development: A review. *Genes. Dev.* **2:** 773.

Levine, M. and T. Hoey. 1988. Homeobox proteins as sequence-specific transcription factors. *Cell* **55:** 537.

Mahon, K.A., H. Westphal, and P. Gruss. 1988. Expression of homeobox gene Hox 1.1 during mouse embryogenesis. *Development* (suppl.) **104:** 187.

Zakany, J., C.K. Tuggle, M.D. Patel, and M.C. Nguyen-Huus. 1988. Spatial regulation of homeobox gene fusions in the embryonic central nervous system of transgenic mice. *Neuron* **1:** 679.

New Approaches to Identifying Genes That Control Early Mammalian Embryogenesis

G.R. Martin, M. Dush, and M.A. Frohman

Department of Anatomy, University of California, San Francisco
School of Medicine, San Francisco, California 94143

In recent years, a number of genes have been identified that are likely to play a role in the control of vertebrate development. Of particular interest is the finding that many of these genes are evolutionarily related and thus belong to families. A striking example is the very large family of homeo-box-containing genes. The existence of such familial relationships can be exploited to isolate previously unknown genes. The homeo box sequence element, for example, has been used by many investigators as a probe to screen genomic or cDNA libraries at reduced stringency to isolate new family members.

All of the mouse homeo-box-containing genes that have been identified in this way are expressed in the developing embryo. Analysis of the available data from in situ RNA hybridization studies has led to the hypothesis that many of these genes participate in the assignment of region-specific positional identity to cells along the rostral-caudal axis during and after the late gastrulation period of development (Holland and Hogan 1988b; Duboule and Dollé 1989; Graham et al. 1989). However, it is unclear what role, if any, members of this gene family play in the control of developmental events immediately prior to or during the early stages of gastrulation, from 6.0 to 7.5 days of development, when the basic body plan of the embryo is first being established. Expression studies of only five homeo-box-containing genes (*Hox-1.1* [Mahon et al. 1988], *Hox-1.5* and *Hox-3.1* [Gaunt 1987, 1988], *Hox-2.1* [Holland and Hogan 1988a], and *Hox-7* [Robert et al. 1989]) have been conducted at these early stages of postimplantation development: Transcripts of these genes are not detected by in situ hybridization prior to 7.5 days of gestation.

Our goal is to develop methods for rapidly isolating and characterizing new members of known gene families that are expressed in the embryo during the early postimplantation

97

stages of development (6.0–7.5 days of gestation). The strategy that we have developed is illustrated in Figure 1. The first step is to isolate embryos at the stages of interest and to extract total cellular RNA from them. Next, the polyadenylated RNAs in the samples are reverse transcribed, using a primer consisting of oligo(dT) (17 residues) linked to a unique oligonucleotide ("adapter") sequence. These steps produce a collection of cDNA molecules. The advantage of this method is that the cDNA collections can be prepared using picogram quantities of total RNA, in contrast with plasmid or phage cDNA libraries, which

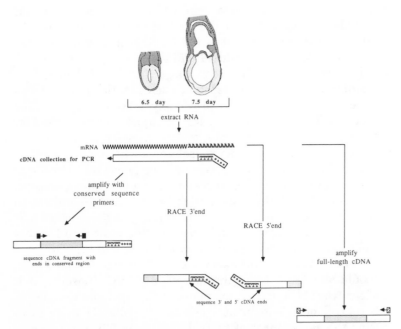

Figure 1 Strategy for rapidly isolating and characterizing new members of known gene families that are expressed in the early postimplantation mouse embryo. Total RNA is extracted from embryos collected at developmental stages of interest, and mRNAs in the samples are reverse transcribed using an oligo(dT) "adapter" primer (TTTT****). Sequences that are conserved in the gene family of interest are used as primers (closed boxes with arrows) in PCR to amplify cDNA segments (stippled boxes), which are then cloned and sequenced to identify new family members. The information thus obtained is used to design gene-specific primers to amplify separately the 3′ and 5′ ends of the cDNAs of interest by the RACE method (Frohman et al. 1988). Sequence analysis of the cDNA ends provides information for designing primers (cross-hatched boxes with arrows) that can be used to obtain full-length cDNA clones by the PCR method.

are constructed from microgram quantities of poly(A)$^+$ RNA. Thus, it is feasible to obtain cDNA collections from tissue samples that contain very small numbers of cells, e.g., a few embryos at very early stages of development. cDNAs in the collection that encode genes in the family of interest are then amplified by the polymerase chain reaction (PCR) method (Saiki et al. 1988). By using as amplification primers the sequences that are conserved in all or most known family members, one can thus amplify the segment of the cDNA that lies between these primers. The amplification products can then be cloned and analyzed further. Clones of known genes in the family or of genes closely related to known family members can be identified by hybridization under appropriate conditions. More distantly related family members can be identified by sequence analysis of unselected clones.

Once a segment of the cDNA is obtained, the remainder can be isolated from the cDNA collection by a method we have developed known as the "rapid amplification of cDNA ends" (RACE) protocol (Frohman et al. 1988). In essence, this method is used to obtain full-length cDNAs by using PCR to amplify copies of the region between a single point in the transcript and the 3′ and 5′ end; once the sequence of each end is known, an additional amplification reaction can be carried out to generate full-length cDNAs (Fig. 1).

Isolation of New Homeo-box-containing Genes
In the course of developing and testing different aspects of this strategy, we have isolated three new mouse homeo-box-containing genes. Two of these are related to the *even-skipped* (*eve*) gene of *Drosophila* (Macdonald et al. 1986; Frasch et al. 1987). They were obtained from a 12.5-day mouse embryo cDNA collection, using as amplification primers a mixture of all the oligonucleotide sequences that could possibly encode one peptide sequence near the amino-terminal end of the *eve* homeo domain and another near the carboxy-terminal end. The amplification products were screened at low stringency with an *eve* homeo box probe, and two different clones were isolated. The RACE protocol is currently being used to isolate full-length cDNA clones of these two genes. Sequence analysis has shown that both genes contain a homeo domain that is almost identical with the homeo domain in the *Drosophila eve* gene. In one case, 56 of 60, and in the other, 55 of 60 amino acids encoded are identical to those found in the *Drosophila* gene. Outside the

homeo domain, however, no sequence similarity between either of the mouse genes and the *Drosophila* gene has yet been detected. Further analysis of these two mouse genes and studies to determine when and where they are expressed in the developing mouse embryo are in progress.

In an early series of pilot experiments, amplification of mouse genomic DNA followed by cloning of the PCR products and screening at reduced stringency with a probe for the *Drosophila zerknüllt (zen)* gene (Rushlow et al. 1987) led to the isolation of three homeo-box-containing gene fragments. One was found to be part of the mouse *Hox-1.6* gene (Baron et al. 1987; LaRosa and Gudas 1988) and a second to be part of the *Hox-5.1* gene (Featherstone et al. 1988). The third contained sequences of homeo box more like the homeo box of the *Drosophila labial* gene (Mlodzik et al. 1988; Diederich et al. 1989) than that of *zen*.

A full-length cDNA clone of this new gene was obtained using the RACE protocol. This gene was mapped by a recombinant inbred strain analysis to the Hox-2 complex on chromosome 11. Sequence comparison has indicated that it is the mouse cognate of the human *HOX-2.9* gene (Boncinelli et al. 1989), and we have therefore designated it *Hox-2.9*. This gene encodes a homeo domain that shares 88% identity with that of another *labial*-like gene, *Hox-1.6* (Baron et al. 1987; LaRosa and Gudas 1988); little sequence conservation between *Hox-2.9* and *Hox-1.6* is present outside the homeo box. A surprising finding is that a short homeo-box-like sequence is located in the single *Hox-2.9* intron. This second sequence is quite different and in the opposite orientation from the *Hox-2.9* homeo box. Experiments to determine whether this sequence is part of a transcription unit are in progress.

Northern blot analysis indicates that *Hox-2.9* expression is strikingly different from that of the other genes in the Hox-2 complex (Graham et al. 1989). Their expression peaks around 14 days of gestation and persists throughout the remainder of embryogenesis, where *Hox-2.9* expression is strong at 9.5–10.5 days of development, decreases at 11.5 days, and is subsequently undetectable. This suggests that the role of *Hox-2.9* in embryogenesis is temporally restricted.

Preliminary in situ RNA hybridization analysis of embryos at 9.5 and 10.5 days of development demonstrates localization of *Hox-2.9* transcripts to the fourth rhombomere of the hindbrain with sharp anterior and posterior limits of expression at the segment boundaries. Such restricted expression has not been reported previously for murine homeo-box-containing genes. We are examining embryos at earlier stages to determine whether the pattern of expression of *Hox-2.9* is consistent with a role in defining the segment boundaries.

We believe that the approach outlined here will prove extremely useful in identifying genes that play a role in early mammalian development. It should enable us to determine which of the known genes in a particular family are expressed at the early postimplantation and preimplantation stages and, more importantly, to isolate new members of the family that are expressed at critical periods in early development, stages that have previously been refractory to analysis because of the limitations in the amount of experimental material available.

REFERENCES

Baron, A., M.S. Featherstone, R.E. Hill, A. Hall, B. Galliot, and D. Duboule. 1987. *Hox-1.6*: A mouse homeo-box containing gene member of the Hox-1 complex. *EMBO J.* **6**: 2977.

Boncinelli, E., D. Acampora, M. Pannese, M. D'Esposito, R. Somma, G. Gaudino, A. Stornaiuolo, M. Cafiero, A. Faiella, and A. Simeone. 1989. Organization of human class I homeobox genes. *Genome* (in press).

Diederich, R.J., V.K.L. Merrill, M.A. Pultz, and T.C. Kaufman. 1989. Isolation, structure, and expression of *labial*, a homeotic gene of the Antennapedia complex involved in *Drosophila* head development. *Genes Dev.* **3**: 399.

Duboule, D. and P. Dollé. 1989. The murine Hox gene network: Its structural and functional organization resembles that of *Drosophila* homeotic genes. *EMBO J.* **8**: 1497.

Featherstone, M.S., A. Baron, S.J. Gaunt, M.-G. Mattei, and D. Duboule. 1988. *Hox-5.1* defines a homeo-gene locus on mouse chromsome 2. *Proc. Natl. Acad. Sci.* **85**: 4760.

Frasch, M., T. Hoey, C. Rushlow, H. Doyle, and M. Levine. 1987. Characterization and localization of the *even-skipped* protein of *Drosophila*. *EMBO J.* **6**: 749.

Frohman, M.A., M.K. Dush, and G.R. Martin. 1988. Rapid production of full-length cDNAs from rare transcripts by amplification using a single gene-specific oligonucleotide primer. *Proc. Natl. Acad. Sci.* **85**: 8998.

Gaunt, S.J. 1987. Homeobox gene *Hox-1.5* expression in mouse embryos: Earliest detection by in situ hybridization is during gastrulation. *Development* **101**: 51.

———. 1988. Mouse homeobox gene transcripts occupy different but overlapping domains in embryonic germ layers and organs: A comparison of *Hox-3.1* and *Hox-1.5*. *Development* **103**: 135.

Graham, A., N. Papalopulu, and R. Krumlauf. 1989. The murine and *Drosophila* homeobox gene complexes have common features of organization and expression. *Cell* **57**: 367.

Holland, P.W. and B.L.M. Hogan. 1988a. Spatially restricted patterns of expression of the homeobox-containing gene Hox-2.1 during mouse embryogenesis. *Development* **102:** 159.

―――. 1988b. Expression of homeo box genes during mouse development: A review. *Genes Dev.* **2:** 773.

LaRosa, G.J. and L.J. Gudas. 1988. Early retinoic acid-induced F9 teratocarcinoma stem cell gene ERA-1: Alternate splicing creates transcripts for a homeobox-containing protein and one lacking the homeobox. *Mol. Cell Biol.* **8:** 3906.

Macdonald, P., P. Ingham, and G. Struhl. 1986. Isolation, structure, and expression of *even-skipped*: A second pair-rule gene of *Drosophila* containing a homeo box. *Cell* **47:** 721.

Mahon, K.A., H. Westphal, and P. Gruss. 1988. Expression of homeobox gene *Hox1.1* during mouse embryogenesis. *Development suppl.* **104:** 187.

Mlodzik, M., A. Fjose, and W.J. Gehring. 1988. Molecular structure and spatial expression of a homeobox gene from the *labial* region of the Antennapedia-complex. *EMBO J.* **7:** 2569.

Robert, B., D. Sassoon, B. Jacq, W. Gehring, and M. Buckingham. 1989. *Hox-7*, a mouse homeobox gene with a novel pattern of expression during embryogenesis. *EMBO J.* **8:** 91.

Rushlow, C., J. Doyle, T. Hoey, and M. Levine. 1987. Molecular characterization of the zerknüllt region of the Antennapedia gene complex in *Drosophila*. *Genes Dev.* **1:** 1268.

Saiki, R.K., D.H. Gelhand, B. Stoffel, S.J. Scharf, R. Higuchi, G.T. Horn, K.B. Mullis, and H.A. Erlich. 1988. Primer-directed enzymatic amplification of DNA with a thermostable DNA polymerase. *Science* **239:** 487.

Using *lacZ* as an In Situ Cell Marker to Analyze Tissue Lineages in the Midgestation Mouse Embryo

R.S.P. Beddington

Imperial Cancer Research Fund Developmental Biology Unit
Department of Zoology, Oxford OX1 3PS, United Kingdom

To determine the effects of specific genes expressed during development, following either mutation of the genes or deliberate ectopic expression, it is necessary to appreciate the normal behavior of cells in the developing embryo. Therefore, the study of cell lineages forms an integral part of any molecular explanation of development. One might anticipate that the mammalian homologs of those invertebrate "developmental" genes shown to play an important part in differentiation, morphogenesis, or pattern formation, are most likely to have an analogous role in mammals in the embryo itself rather than in the more recently evolved extraembryonic constituents of the conceptus. However, in the mouse the pattern of development of extraembryonic tissues, particularly primitive endoderm derivatives, is better understood than the deployment of cells during the earliest stages of embryonic organization (Gardner 1985). There are two main reasons for this: (1) The preimplantation embryo, which is primarily concerned with segregating extraembryonic tissue lineages, is readily accessible to manipulation, whereas the early postimplantation conceptus, during gastrulation and early organogenesis, is not. (2) Until recently, no in situ cell markers have been available that could be used to study cell or tissue lineages in the epiblast, the precursor of the entire fetus, and its immediate derivatives.

Recently, a transgenic mouse line carrying a bacterial expression marker, whose protein product can be detected in all cells of the embryo at midgestation, has been made (Beddington et al. 1989). Chimeras, resulting either from injecting a single transgenic inner cell mass (ICM) cell into wild-type blastocysts or from transplanting a single transgenic somite into postimplantation wild-type embryos, demonstrate how such a marker provides information regarding not only cell or tissue fate, but also reveals complex patterns of growth and cell intermixing.

Transgenic Mouse Line *Tg(Act-lacZ)-1*

The transgenic mouse line *Tg(Act-lacZ)-1* (Beddington et al. 1989) contains a concatamer composed of six copies of the 4.3-kb construct illustrated in Figure 1. No homozygous transgenics have yet been identified, and therefore it is assumed that homozygosity for the integration is lethal to embryos. Furthermore, there is some evidence that the transgene may not be neutral to gametes.

Staining for bacterial β-galactosidase activity in hemizygous conceptuses during development reveals that activity is first detectable at the onset of gastrulation and is restricted to epiblast cells and their derivatives. On the tenth day of gestation, all cells in the embryo show generalized cytoplasmic staining, and the extraembryonic mesoderm exhibits patchy staining. Evidence from double-labeling experiments, in which ninth day [³H]thymidine-labeled transgenic tissue was grafted into wild-type embryos of the same age, which were subsequently allowed to develop in culture for 24 hours, indicates that this staining is cell autonomous. The trophectoderm and primitive endoderm derivatives are completely negative throughout gestation.

On the eleventh day of gestation, the first signs of restricted expression in the embryo are apparent. Progressive reduction in the expression of *lacZ* continues until birth, and within 4 or 5 weeks postnatally very few cells contain detectable bacterial β-galactosidase activity. The loss of β-galactosidase activity appears to occur not only in specific tissues, but also in particular domains within certain tissues (P. Savatier, unpubl.), reminiscent of the chromosomal position effects associated with integration site reported for other transgenic lines (Allen et al.

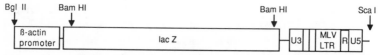

Figure 1 Diagram of the 4.2-kb DNA construct used to produce the *Tg(Act-lacZ)-1* mouse line following pronuclear injection. The β-actin promoter is rat in origin. (MLV LTR) Moloney leukemia virus 3′ long terminal repeat.

1988). Superimposed on this pattern of negative activity is an apparently random loss of activity among cells of the same tissue, indicating a more generalized and sustained down-regulation.

ICM Clonal Analysis

The ubiquitous and cell-autonomous expression of *lacZ* in the tenth day embryo means that the distribution of ICM clones can be assessed for the first time in situ during the early stages of organogenesis. Analysis of chimeras derived from injecting a single hemizygous *Tg(Act-LacZ)-1* ICM into wild-type blastocysts confirms that ICM cells are indeed pluripotent and invariably colonize all the major epiblast derivatives present at midgestation (Beddington et al. 1989). Furthermore, clonal descendants are present in all regions of the embryo extending from the most anterior to the most posterior extremes of the body axis. The high levels of chimerism observed and the widespread distribution of transgenic cells, when in competition with wild-type ones, argue that *lacZ* expression is neutral to development at least during the early stages of postimplantation development.

The chimeras also provide dramatic evidence for extensive cell mixing, visible both in whole-mount preparations and in sectioned material. By the tenth day of gestation, there is some indication of limited coherent growth in certain organ primordia, such as the gut, neural tube, and individual somites. All somites were derived from both genotypes, which confirms previous conclusions that they are not clonal in origin (Gearhart and Mintz 1972).

Somite Transplantation

The widespread distribution of ICM clones found at the onset of organogenesis means that preimplantation manipulation of the embryo is of limited value in determining the segregation of tissue lineages within the embryo itself. Consequently, marked cells must be introduced after implantation in order to ascertain the ancestry of particular fetal primordia. Similarly, the lability of cells during organogenesis, which is of paramount importance in assessing the relative importance of particular patterns of gene expression, can only be examined by perturbing the postimplantation embryo. Preliminary results from transplanting transgenic somites into wild-type hosts, which were subsequently cultured for 48 hours, illustrates the potential of postimplantation chimeras for resolving tissue lineages.

Replacing a somite, destined to be opposite the forelimb bud, in a wild-type embryo with a single *Tg(Act-lacZ)-1* somite demonstrates the diversity and chronology of the somitic contribution of the limb. The extensive invasion of the limb bud by endothelial cells expressing bacterial β-galactosidase suggests that in mammals somites may give rise to vascular elements. The provision of limb musculature by somites has yet to be unequivocally demonstrated in the mouse, but it would appear that both the migration of myogenic cells and the formation of muscle blocks in the limb occur at a relatively later stage of limb development (R. Beddington and P. Martin, unpubl.) than that seen in the chick (Chevallier et al. 1977; Christ et al. 1977).

The demonstration that transplanted somites readily incorporate into host embryos and apparently differentiate in parallel with adjacent host somites, forming characteristic myotome and sclerotome tissue, also provides an opportunity for examining the regulation of axial-specific patterns of gene expression. For example, the cell autonomy, or otherwise, of the expression of particular *Hox* genes could be examined following heterotopic transplantation of particular somites.

PROSPECTS

Early mouse development seems to be characterized by a high capacity for regulation and compensation following perturbation. This suggests that a molecular and genetic dissection of mammalian development may be less straightforward than in lower organisms where cell-lineage divergence is more regular and invariant. It may therefore be necessary to devise more subtle assays in the mammalian embryo to study the effects of particular genetic modifications. By harnessing efficient, ubiquitous, in situ cell markers to appropriate chimera studies, it should be possible to establish some of the basic cellular rules governing gastrulation and early organogenesis. Subsequently, the behavior of similarly marked but genetically modified embryonic (or embryonic stem) cells in chimeras can be properly evaluated.

ACKNOWLEDGMENTS

I thank Dr. Karen Downs and Dr. Pierre Savatier for useful discussion of the manuscript and Dr. Paul Martin for allowing inclusion of unpublished results. This work was supported by the Imperial Cancer Research Fund.

REFERENCES

Allen, N.D., D.G. Cran, S.C. Barton, S. Hettle, W. Reik, and M.A. Surani. 1988. Transgenes as probes for active chromosomal domains in mouse development. *Nature* **333**: 852.

Beddington, R.S.P., J. Morgernstern, H. Land, and A. Hogan. 1989. An *in situ* transgenic enzyme marker for the midgestation mouse embryo and the visualization of inner cell mass clones during early organogenesis. *Development* **106**: 37.

Chevallier, A., M. Kieny, and A. Mauger. 1977. Limb-somite relationship: Origin of the limb musculature. *J. Embryol. Exp. Morphol.* **41**: 245.

Christ, B., H.J. Jacob, and M. Jacob. 1977. Experimental analysis of the origin of the wing musculature in avian embryos. *Anat. Embryol.* **150**: 171.

Gardner, R.L. 1985. Clonal analysis of early mammalian development. *Proc. R. Soc. Lond. B Biol. Sci.* **312**: 163.

Gearhart, J.D. and B. Mintz. 1972. Clonal analysis of somites and their muscle derivatives: Evidence from allophenic mice. *Dev. Biol.* **29**: 27.

Developmental Expression of the Murine *vgr-1* Gene: A New Member of the TGF-β Multigene Family

K.M. Lyons,[1] R.W. Pelton,[1] J.L. Graycar,[2] R. Derynck,[2] and B.L.M. Hogan[1]

[1]Department of Cell Biology, Vanderbilt University School of Medicine, Nashville, Tennessee 37232

[2]Department of Developmental Biology, Genentech, Inc. South San Francisco, California 94080

It has become increasingly clear that growth factors and their receptors are involved in the control of cellular proliferation, differentiation, and normal embryonic development (Adamson 1987; Hanley 1988; Mercola and Stiles 1988). In this context, one particularly interesting class of growth factors is the transforming growth factor-β (TGF-β) superfamily. This superfamily contains a group of at least five gene products that are closely related to TGF-β, the first member to be characterized, as well as to a number of more distantly related molecules (e.g., Müllerian inhibiting substance and the *Drosophila decapentaplegic* [*dpp*] gene product). TGF-β1 and TGF-β2 have been isolated from a variety of adult tissues and can either stimulate or inhibit cell proliferation and differentiation in model systems in vitro, (for example, see Silberstein and Daniel 1987; Coffey et al. 1988; Rosen et al. 1988). In addition, immunohistochemical (Heine et al. 1987), in situ hybridization (Lehnert and Akhurst 1988; Wilcox and Derynck 1988; Pelton et al. 1989), and Northern blot analyses (D.A. Miller et al., in prep.) have shown that TGF-β1, -β2, and -β3 are expressed in specific tissues during embryonic development in the mouse, suggesting that these genes are involved in early development, as well as in the regulation of growth and differentiation of adult tissues. The demonstration that TGF-β2 alone (Rosa et al. 1988) or TGF-β1 in combination with basic fibroblast growth factor (Kimmelman and Kirschner 1987) induces the expression of a mesoderm-specific marker gene in isolated *Xenopus* animal cap cells provides additional evidence for a role for TGF-β1 and -β2 in early development.

The members of a second more distantly related group of TGF-β-like genes show greater homology with the *Drosophila dpp* gene product than with TGF-β1. This subgroup of *dpp*-like gene products includes the bone morphogenetic proteins (BMPs) types-2a, -2b, and -3 (Wozney et al. 1988) and the *Xenopus Vg-1* gene (Weeks and Melton 1987). Direct evidence for a role for *dpp* in embryonic development is provided by genetic studies that show that embryos homozygous for null alleles of *dpp* are completely ventralized (Irish and Gelbart 1987). Furthermore, transcripts of the *dpp* gene are detected in the embryo and larva predominantly in the dorsal ectoderm, but also in the visceral mesoderm and gut endoderm (St. Johnston and Gelbart 1987). *Vg-1* was isolated as a maternally encoded mRNA that becomes progressively localized to the presumptive endoderm (vegetal hemisphere) of *Xenopus* embryos and then declines after gastrulation (Rebagliati et al. 1985). Signals originating in the vegetal hemisphere are thought to induce mesoderm formation in the overlying ectoderm in a complex process involving several different signals (Dale and Slack 1987; Smith 1989). On the basis of its temporal and spatial expression, *Vg-1* is one candidate for a mesoderm-inducing factor.

These data show that TGF-β-like molecules are present in phylogenetically distant organisms, and they are likely to be involved in some aspect of the control of growth and differentiation. We reasoned that genes more closely related to the *Drosophila dpp* and *Xenopus Vg-1* genes than to TGF-β1 might be present in the mouse and that such genes may play a role in pattern formation and normal cell growth. Therefore, we screened an 8.5-day postcoitus mouse embryo cDNA library with a probe derived from the *Xenopus Vg-1* gene, and we isolated a cDNA designated *vgr-1* (vg-related). The protein encoded by the *vgr-1* cDNA contains a carboxy-terminal sequence that includes seven cysteine residues that are highly conserved among all of the members of the TGF-β superfamily. This region shows 59%, 57%, 61%, 59%, and 44% identity with the corresponding regions of *Vg-1, dpp,* and BMPs-2a, -2b, and -3. In contrast, *vgr-1* has only 31% identity with the corresponding region of TGF-β1 (Lyons et al. 1989). *vgr-1* thus belongs to a distinct subfamily of *dpp*-like genes within the TGF-β superfamily, and it is the first member of this subfamily to be described in the mouse.

The pattern of *vgr-1* expression was assayed by Northern

blot analysis of RNA isolated from embryonic, neonatal, and adult tissues. In many tissues, the level of a single 3.5-kb transcript increases throughout development and into adulthood. For example, the level increases in the kidney and lung during embryogenesis and continues to increase after birth. A relatively smaller increase was seen in calvaria and skin. In these tissues, a single 3.5-kb species of *vgr-1* transcript is detected. In the testes, an additional 1.8-kb transcript is present in approximately equal proportion to the 3.5-kb mRNA. Poly(A)$^+$ RNA from a variety of normal and neoplastic cell lines known to express TGF-β1, -β2, or -β3 was also surveyed. Unlike the TGF-β types, which are expressed in many cell lines, *vgr-1* expression was detected in only three of the cell lines known to express TGF-β, normal rat kidney fibroblasts (NRK-49F), the human pancreatic cell line panc-1, and the human fibrosarcoma cell line HT1080 (Lyons et al. 1989). Taken together, these data suggest that, like the *Drosophila dpp* gene (Irish and Gelbart 1987; St. Johnston and Gelbart 1987), *vgr-1* plays a role at several stages of development.

We have used in situ hybridization to determine more precisely the pattern of *vgr-1* expression. This analysis has revealed that *vgr-1* is expressed in many of the same tissues as TGF-β1 (Lehnert and Akhurst 1988; Sandberg et al. 1988) and TGF-β2 (Pelton et al. 1989) but in different cell types. For example, in developing limbs *vgr-1* transcripts are restricted to the hypertrophic cells of the cartilage, whereas TGF-β1 and TGF-β2 are found in the osteoblasts, osteocytes, and osteoclasts of developing bone. TGF-β2 is also expressed in the perichondrium but not in the hypertrophic cartilage. In developing skin (Fig. 1A–C), *vgr-1* is expressed at high levels in the suprabasal layer of the proliferating epidermis. Expression of *vgr-1* mRNA is first detected in this layer at day 15.5 postcoitus, when the epidermis begins to thicken, and continues throughout embryogenesis and after birth. No *vgr-1* expression is detected in the dermis or hair follicles at any stage of development. In contrast, TGF-β2 is localized to the dermis at day 15.5 postcoitus (Fig. 1D–F). On day 17.5 postcoitus, TGF-β2 expression in the dermis declines, and by day 18.5 postcoitus TGF-β2 mRNA appears in the suprabasal layer of the epidermis and in the hair follicles. This pattern of expression persists throughout the remainder of embryogenesis and is still seen at least as late as 3 days after birth (Fig. 1G–I).

These related but clearly distinct temporal and spatial pat-

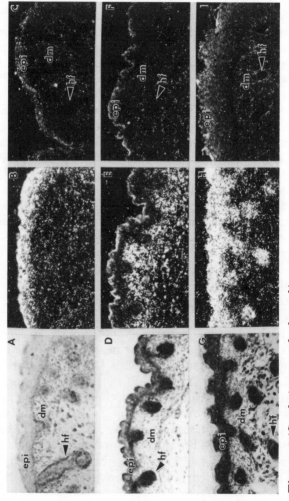

Figure 1 (*See facing page for legend.*)

112

terns of expression of TGF-β-like genes suggest that they participate in the development of many of the same tissues but have different roles. Thus, overlapping and/or complementary gradients of expression of TGF-β-like gene products might be required for the coordination of growth and differentiation in tissues where cellular interactions occur. For example, normal limb development requires the coordinated differentiation of chondrocytes into hypertrophic cartilage and the migration of osteoblasts into the zone of calcifying cartilage and their differentiation into osteocytes. TGF-β-like gene products may influence the level of their own expression in an autocrine manner (for example, see Coffey et al. 1988) as well as that of other TGF-β-like genes in a paracrine manner, thereby coordinating the growth and differentiation of different cell types within a single developing system. An additional level of coordination might be achieved at the posttranslational level, as exemplified by the activins and inhibins. For example, activin, which stimulates follicle-stimulating hormone synthesis and release in the pituitary gland, is a homodimer composed of two TGF-β-like β-subunits linked by disulfide bonds. Inhibin, on the other hand, has the opposite effect and is a heterodimer composed of one α-subunit and one β-subunit (Ling et al. 1986; Vale et al.

Figure 1 Expression of *vgr-1* and TGF-β2 mRNA in developing skin. (A) Section through the skin of a 16.5-day postcoitus embryo hybridized with an antisense *vgr-1* probe and photographed under bright-field illumination. (B) Same section as in A, photographed under dark-ground illumination. Strong hybridization is seen in a suprabasal layer of the epidermis but is absent in the dermis and hair follicles. (C) Adjacent section to A hybridized with the sense strand and photographed under dark-ground illumination. The apparent signal in the outermost layer of the epidermis is an artifact due to light scattering by highly keratinized cells. (D) Section through the skin of a 15.5-day postcoitus embryo hybridized with antisense TGF-β2 probe and photographed under bright-field illumination. (E) Same section as in D photographed under dark-ground illumination. Strong hybridization is seen in the dermis but is absent in the epidermis and hair follicles. (F) Adjacent section to D probed with sense probe and photographed under dark-ground illumination. (G) Section through the skin of an 18.5-day postcoitus embryo hybridized with antisense TGF-β2 probe and photographed under bright-field illumination. (H) Same section as in G photographed under dark-ground illumination. Strong signal is evident in the epidermis and hair follicles but is not seen in the dermis. (I) Adjacent section to G hybridized to the TGF-β2 sense probe and photographed under dark-ground illumination. Sections were exposed 5–15 days. (epi) Epidermis; (dm) dermis; and (hf) hair follicle.

1986). Thus, heterodimers composed of different TGF-β-like gene products might be expected to have activities that are very different from the corresponding homodimers.

In addition to our studies of *vgr-1* localization in late-stage embryos, we have been examining *vgr-1* expression at early stages of development. In the ovary, *vgr-1* mRNA is expressed at high levels in developing and mature oocytes (Fig. 2). We have also studied *vgr-1* expression in embryonal carcinoma F9 cells. In response to retinoic acid, these cells can be induced to differentiate into either parietal endoderm (in monolayer) or embryoid bodies with an outer layer of endoderm (in suspension). We have shown by Northern blot analysis that *vgr-1* expression is strongly induced by differentiation of F9 cells into either parietal endoderm or embryoid bodies in response to retinoic acid. Endodermally derived parietal yolk sac cells also express *vgr-1* (Lyons et al. 1989). This pattern of expression is comparable with the localization of the *Vg-1* mRNA to the vegetal pole (presumptive endoderm) in *Xenopus* embryos (Melton 1987). By analogy with *Xenopus* embryogenesis, it might be expected that the first endoderm to appear in the mouse may contain factors that are involved in the initial formation of meso-

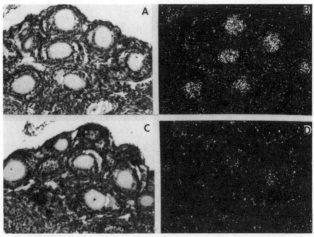

Figure 2 Expression of *vgr-1* in adult ovary. (A) Section through an adult ovary hybridized with an antisense *vgr-1* probe and photographed under bright-field illumination. (B) Same section as in A photographed under dark-ground illumination. Strong hybridization is seen in oocytes. (C) Adjacent section to A hybridized with sense probe and photographed under dark-ground illumination.

derm. The murine *vgr-1* gene may therefore be a candidate for one of the mesoderm-inducing factors.

In future work, we plan to explore further the possibility that *vgr-1* plays a role in morphogenetic events in the mouse. The production of transgenic mice will allow us to examine the effects of inappropriate temporal and/or spatial *vgr-1* expression in specific developing systems such as cartilage and skin. The generation of homozygous null mutants for *vgr-1* by gene targeting will provide genetic evidence concerning the role of *vgr-1* during development and the nature of the interactions that occur between the *vgr-1* gene product and other TGF-β-like molecules. Finally, the availability of antibodies and biologically active *vgr-1* protein and identification of the *vgr-1* receptor are key requirements for a more precise definition of the role this gene plays at early and later stages of murine development.

ACKNOWLEDGMENTS

We thank Sofie Hashmi for excellent technical assistance and Dr. Harold Moses for generous support. K.M.L was supported in part by Vanderbilt University National Cancer Institute institutional postdoctoral grant 5T32-CA-0959292. R.W.P. was supported by National Institutes of Health grant T32-GM-07347 for the medical scientist training program.

REFERENCES

Adamson, E.D. 1987. Oncogenes in development. *Development* **99**: 449.

Coffey, R.J., N.J. Sipes, C.C. Bascom, R. Graves-Deal, C.Y. Pennington, B.E. Weissman, and H.L. Moses. 1988. Growth modulation of mouse keratinocytes by transforming growth factors. *Cancer Res.* **48**: 1596.

Dale, L. and M.W. Slack. 1987. Regional specification within the mesoderm of early embryos of *Xenopus laevis*. *Development* **100**: 279.

Hanley, M.R. 1988. Proto-oncogenes in the nervous system. *Neuron* **1**: 175.

Heine, U.I., E.F. Munoz, K.C. Flanders, L.R. Ellingsworth, H.-Y.P. Lam, N.L. Thompson, A.B. Roberts, and M.B. Sporn. 1987. Role of transforming growth factor-β in the development of the mouse embryo. *J. Cell Biol.* **105**: 2861.

Irish, V.F. and W.M. Gelbart. 1987. The *decapentaplegic* gene is required for dorsal-ventral patterning of the *Drosophila* embryo. *Genes Dev.* **1**: 868.

Kimmelman, D. and M. Kirschner. 1987. Synergistic induction of mesoderm by FGF and TGF-β and the identification of an mRNA coding for FGF in the early *Xenopus* embryo. *Cell* **51**: 869.

Lehnert, S.A. and R.J. Akhurst. 1988. Embryonic expression pattern of TGF beta type-1 RNA suggests both paracrine and autocrine mechanisms of action. *Development* **104**: 263.

Ling, N., S.-Y. Ying, N. Ueno, S. Shimasaki, F. Esch, M. Hotta, and R. Guillemin. 1986. Pituitary FSH is released by a heterodimer of the β-subunits from the two forms of inhibin. *Nature* **321**: 779.

Lyons, K.M., J.L. Graycar, A. Lee, S. Hashmi, P.B. Lindquist, E.Y. Chen, B.L.M. Hogan, and R. Derynck. 1989. Vgr-1, a mammalian gene related to *Xenopus* vg-1 and a new member of the TGF-β gene superfamily. *Proc. Natl. Acad. Sci.* **86**: 4554.

Melton, D.A. 1987. Translocation of a localized maternal mRNA to the vegetal pole of *Xenopus* oocytes. *Nature* **328**: 80.

Mercola, M. and C.D. Stiles. 1988. Growth factor superfamilies and mammalian embryogenesis. *Development* **120**: 451.

Pelton, R.W., S. Nomura, H.L. Moses, and B.L.M. Hogan. 1989. Expression of transforming growth factor β2 RNA during murine embryogenesis. *Development* (in press).

Rebagliati, M.R., D.L. Weeks, R.P. Harvey, and D.A. Melton. 1985. Identification and cloning of localized maternal RNAs from *Xenopus* eggs. *Cell* **42**: 769.

Rosa, F., A.B. Roberts, D. Danielpour, L.L. Dart, D.B. Sporn, and I.B. Dawid. 1988. Mesoderm induction in amphibians: The role of TGF-β2-like factors. *Science* **239**: 783.

Rosen, D.M., S.A. Stempien, A.Y. Thompson, and S.M. Seyedin. 1988. Transforming growth factor-beta modulates expression of osteoblast and chondroblast phenotypes in vitro. *J. Cell. Physiol.* **134**: 337.

Sandberg, M., T. Vourio, H. Hirvonen, K. Alitalo, and E. Vourio. 1988. Enhanced expression of TGF-β and c-*fos* mRNAs in the growth plates of developing human long bones. *Development* **102**: 461.

Silbertstein, G.B. and C.W. Daniel. 1987. Reversible inhibition of mammary gland growth by transforming growth factor-β. *Science* **237**: 291.

Smith, J.C. 1989. Mesoderm induction and mesoderm-inducing factors in early amphibian development. *Development* **105**: 665.

St. Johnston, R.D. and W.M. Gelbart. 1987. Decapentaplegic transcripts are localized along the dorsal-ventral axis of the *Drosophila* embryo. *EMBO J.* **6**: 2785.

Vale, W., J. Rivier, J. Vaughan, R. McClintock, A. Corrigan, W. Woo, D. Karr, and J. Spiess. 1986. Purification and characterization of an FSH releasing protein from porcine ovarian follicular fluid. *Nature* **321**: 776.

Weeks, D.L. and D.A. Melton. 1987. A maternal mRNA localized to the vegetal hemisphere in *Xenopus* eggs codes for a growth factor related to TGF-β. *Cell* **51**: 861.

Wilcox, J.N. and R. Derynck. 1988. Developmental expression of transforming growth factors alpha and beta in mouse fetuses. *Mol. Cell. Biol.* **8**: 3415.

Wozney, J.M., V. Rosen, A.J. Celeste, L.M. Mitsock, M.J. Whitters, R.W. Kriz, R.M. Hewick, and E.A. Wang. 1988. Novel regulators of bone formation: Molecular clones and activities. *Science* **242**: 1528.

In Situ Analysis of c-*fos* Expression in Transgenic Mice

K. Heckl and E.F. Wagner

Research Institute of Molecular Pathology, A-1030, Vienna, Austria

The proto-oncogene c-*fos* has been the subject of extensive analysis, and the results from various experimental approaches suggest that the c-*fos* protein regulates many different biological processes (for review, see Müller 1986). The pleiotropic action of c-*fos* is suggested by its expression pattern. Expression of c-*fos* has been demonstrated in extraembryonic tissues of the mouse (Müller 1986), in macrophages and neutrophils (Gonda and Metcalf 1984; Kreipe et al. 1987), and in the growth plates of fetal bones (Dony and Gruss 1988; Sandberg et al. 1988) in both mice and humans. In vitro studies have shown that c-*fos* expression can be induced by various growth factors (Müller 1986; Verma and Sassone-Corsi 1987) and that it is required for cell proliferation (Holt et al. 1986; Nishikura and Murray 1987). Recent studies demonstrated that *fos* and *jun* proteins are part of a transcription complex (Curran and Franza 1988) recognizing the AP-1 binding site, which is found within the regulatory regions of certain genes (Chiu et al. 1988; Rauscher et al. 1988; Sassone-Corsi et al. 1988). These observations support the assumption that the *fos* protein complex can modulate the expression of a variety of target genes.

To study the biological function of c-*fos* in vivo, several transgenic mouse lines have been established in our laboratory (Wagner et al. 1989). One series of experiments utilized a DNA vector in which the entire murine c-*fos* gene had been fused to the inducible human metallothionein promoter (MT) and part of the 3′ untranslated region of the c-*fos* gene had been replaced by the Finkel-Biskis-Jinkins sarcoma virus long terminal repeat (LTR) (Rüther et al. 1985). Several transgenic mouse lines have been established, two of which express the transgene at high levels in pancreas, heart, kidney, muscle, brain, and bone tissues of adult mice (Rüther et al. 1987). However, phenotypic changes are observed exclusively in the bones of these animals: 15% of the MT-c-*fos*-LTR mice develop characteristic swellings of the long bones as early as 2–3 weeks after birth. The bone lesions display a marked disturbance of bone remodeling characterized by bone marrow fibrosis and enhanced formation of new bone. The lesions do not increase in size significantly beyond 4 weeks of age. However, 18% of the transgenic animals develop tumors of the long bones, on average, 9 months after birth. These tumors have been classified as fibrosarcomas, osteosarcomas, and chondrosarcomas (Rüther et al. 1989). The specificity of bone tumors seems to be indepen-

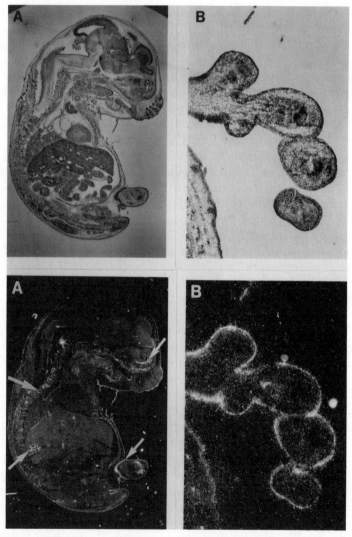

Figure 1 (*See facing page for parts C and D and legend.*)

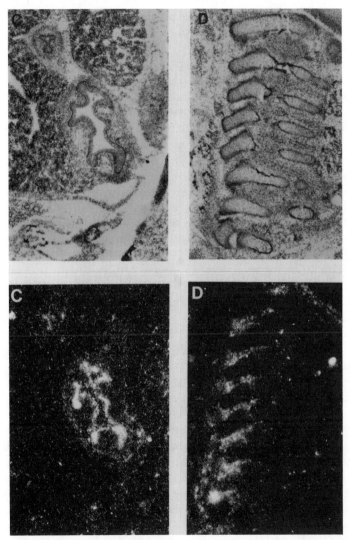

Figure 1 In situ hybridization of median cryosections through a C3H mouse embryo (E17.5) in bright-field and dark-field illumination. (A) Entire embryo. Expressing tissues are indicated by arrows: mesodermal web tissue, developing cartilage (nasal cavity), mucosa, and Schwann cells of the peripheral nerves. (B) Cross-section of hindlimb showing *fos* expression in the mesodermal web tissue. (C) Mucosa. (D) Schwann cells. Probe: Antisense *fos*.

119

dent of the chosen promoter (an H2-c-*fos*-LTR vector yields even more dramatic bone lesions) but dependent on the presence of the LTR (Rüther et al. 1989). It is interesting to note that stable expression of H2-c-*fos* without the LTR in transgenic mice leads to hyperplasia of the thymus (Rüther et al. 1988).

To investigate the pathogenesis of the transgenic mice, we examined the temporal and spatial expression of c-*fos* in the MT-c-*fos*-LTR mice by in situ hybridization. In this paper, we describe the results of the analysis of c-*fos* expression in various tissues, in particular the bones of transgenic mice and control animals, during development and in early postnatal life as well as in the bone tumors.

RESULTS

SP6 ^{35}S-labeled probes were prepared for in situ analysis, using a 1-kb *Pst*I fragment of the murine c-*fos* cDNA (Dony and Gruss 1988), which recognizes endogenous and exogenous c-*fos*. This probe allowed us to analyze the nontransgenic mice as well.

The following mouse strains were analyzed: two transgenic lines that express the MT-c-*fos*-LTR construct, nos. 209-4 and 211-5 of the C3H strain, and control animals, C3H and C57BL/6. In situ hybridization was performed on cryosections of midgestation (E13.5 and E14.5) and late gestation embryos (E17.5). Sections of entire 2-day-old mice (P2) were also analyzed. In parallel, in situ hybridization was performed on normal and transgenic limbs of E17.5 embryos and P2 and P8 animals (same strains as mentioned above).

c-*fos* Expression in Nontransgenic Mice

To assess endogenous c-*fos* expression, in situ hybridization was performed on nontransgenic mice. Expression was detected in E13.5 and E14.5 embryos in the mesodermal web tissue of the digits, in the growth regions of developing bones, and in the developing cartilage (data not shown), thus confirming previous reports (De Togni et al. 1988; Dony and Gruss 1988; Sandberg et al. 1988). At late gestation (Fig. 1A, E17.5), high levels of c-*fos* message were found again in the mesodermal web tissue (Fig. 1B), in the intestine (Fig. 1C, mucosa), and in the developing cartilage of the nasal cavity (Fig. 1A). Moderate levels were detected in muscle (e.g., Fig. 1A, tongue), and in the growth plates of developing long bones (data not shown). Expression in the nervous system was confined to the motor neurons of the anterior horn of the spinal cord (Fig. 1A) and to Schwann cells of the developing peripheral nervous system (Fig. 1D), confirmed by immunocytochemistry (data not shown). Other cell types of the central and peripheral nervous system did not exhibit detectable levels of c-*fos*. The nontransgenic P2 mice did not show any variation of the above mentioned c-*fos* expression pattern: Moderate levels were found in intestine, muscle, cartilage (data not shown), and in regions of bone remodeling (Fig. 3A1,2).

c-*fos* Expression in Transgenic Mice

In transgenic E13.5 embryos, no elevated levels of c-*fos* mRNA could be detected (data not shown). However, high levels, as shown in Figure 2, could be found throughout late gestation and at 2 days after birth: E17.5 transgenic embryos (Fig. 2) and P2 animals (data not shown) of both transgenic lines express c-

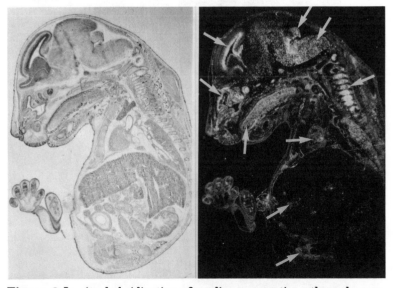

Figure 2 In situ hybridization of median cryosections through a no. 209-4 transgenic mouse embryo (E17.5) in bright-field and dark-field illumination. Expressing tissues are indicated by arrows: undifferentiated precursor in the subependymal germinal plate, astrocytes (throughout the whole brain and spinal cord), cerebellum, sensory neurons of the dorsal root ganglia, nose, muscle (tongue and limbs), heart, and intestine. Probe same as in Fig. 1.

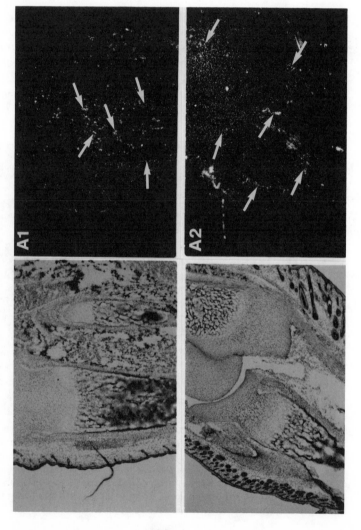

122

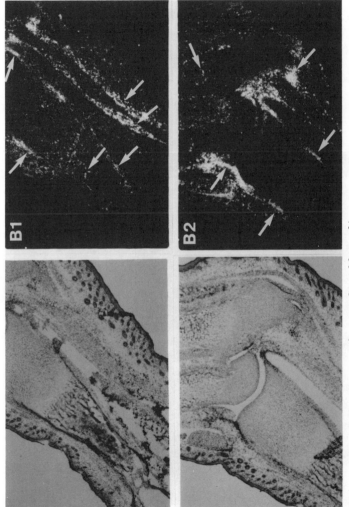

Figure 3 *(See following page for part C and legend.)*

123

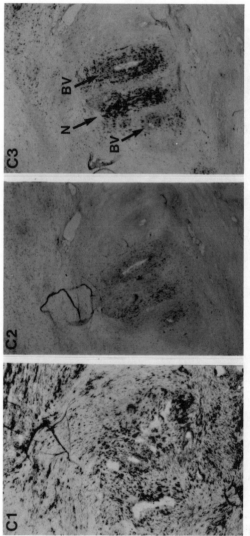

Figure 3 In situ hybridization of longitudinal sections through tibias and ankles of C3H nontransgenic, no. 211-5 transgenic mice at stage P2 and P8, and cross-section of a tumor in bright-field and dark-field illumination. (*A1*) Tibia (P2) and (*A2*) ankle (P8) of a C3H animal. Expressing tissues and regions are indicated by arrows: epiphyseal cartilage, perichondrium, region of enchondrial ossification. Probe same as in Fig. 1. (*B1*) Tibia (P2) and (*B2*) ankle (P2) of a no. 211-5 transgenic animal. Expressing tissues and regions are indicated: epiphyseal cartilage, perichondrium, region of enchondrial ossification, tendons. Probe is same as in Fig. 1. (*C1-3*) Fibrosarcoma, cross-section of the right hindlimb of a no. 209-4 transgenic animal at the age of 12 months. Probe (*C1* and *C3*) antisense *fos* and (*C2*) sense *fos*. Expression in different cell types, e.g., blood vessels (BV), and neurons (N).

124

fos in the central nervous system (subependymal germinal plate and astrocytes of the telencephalon, cerebellum, and spinal cord, and motor neurons in the anterior horn of the spinal grey matter), in the peripheral nervous system (sensory neurons of the dorsal root and trigeminal ganglia, Schwann cells), in the adrenal medulla, as well as in the retina. Elevated levels of the c-*fos* transcript were present in intestine and muscle (heart) as well as in cartilage (nasal cavity and epiphyseal cartilage), in tendons, and in regions of perichondrial ossification (Fig. 2 and Fig. 3B1,2). This analysis largely confirms the RNA expression data obtained in adult mice (Rüther et al. 1987).

c-*fos* Expression in Bone Tissues and Bone Tumors

The long bones and joints, sites of pathological alterations in transgenic animals, were closely investigated in order to identify the cell types exhibiting c-*fos* expression in the affected bones and limbs of transgenic mice and in the bone tumors. Analysis was first carried out on tibias and femurs of E17.5, P2, and P8 limbs of nontransgenic mice. c-*fos* expression was found in regions of perichondrial and enchondrial ossification, as well as in the epiphyseal cartilage (Fig. 3A2, chondroblasts). Since developing bones of mice of the investigated stages do contain only very few osteoclasts and because of other criteria, we would identify the expressing cells within the region of enchondrial ossification as osteoblasts. This contrasts with a report by Sandberg et al. (1988), confining the detected c-*fos* expression within the regions of enchondrial ossification to osteoclasts. In addition, the bone marrow contained small numbers of distinct c-*fos*-expressing cells with no identifiable morphology (data not shown).

Corresponding sections through limbs of both transgenic mouse strains (again E17.5, P2, and P8) showed elevated c-*fos* expression levels. High levels could be found in the tendons, epiphyseal cartilage, and partly in the regions where perichondrial ossification occurs (Fig. 3B2). However, there is no significant difference to control animals in the region of enchondrial ossification.

Analysis of a bone tumor, characterized as a fibrosarcoma (Fig. 3C) that contained fibroblasts, fat cells, cells involved in blood vessel formation, neurons, newly formed bone tissue, and cartilage, showed c-*fos* expression in a subpopulation of all these cells, as well as in unidentified cell types. This confirms our RNA data of tumor tissues, which consistently were found to express about tenfold higher levels of exogenous *fos* RNA when compared with unaffected bone tissue (Rüther et al. 1989).

DISCUSSION

Using in situ hybridization, we have analyzed the pattern of c-*fos* expression in transgenic and nontransgenic mice at embryonic stage E13.5 and E17.5 and at postnatal stage P2. Our analysis has revealed c-*fos* expression in mesodermal web tissue, in the developing cartilage, and in the growth plates of developing long bones. These findings are largely consistent with previous reports (De Togni et al. 1988; Dony and Gruss 1988). Furthermore, we have identified c-*fos* expression at moderate and low levels in muscles (tongue, heart, and limbs) and in particular in the intestine (mucosa). In the nervous system, c-*fos* expression could only be detected in the motor neurons of the anterior horn of the spinal cord and in the Schwann cells of the developing peripheral nervous system. In contrast with a recently published report (Caubet 1989), endogenous c-*fos* could not be detected in any other part of the central nervous system, such as in the spinal ganglia or in the retina.

RNase protection assay of both transgenic mouse lines had revealed high levels of exogenous c-*fos* expression in certain tissues (Rüther et al. 1987). Our investigations clearly demonstrate that this observed pattern of c-*fos* expression in adult transgenics is already detectable as early as late gestation (E17.5) and is sustained during the early postnatal period. In situ analysis combined with immunocytochemistry revealed that the transgene is efficiently expressed in neuronal and glial cells of the central nervous system (undifferentiated precursor of the subependymal germinal layer, as well as in astroglia throughout the whole brain, spinal cord, and motor neurons of the anterior horn of the spinal grey matter) and of the peripheral nervous system (sensory neurons of the dorsal root and trigeminal ganglia, Schwann cells of the peripheral nerves, in the retina, and in the adrenal medulla). Despite high levels of ectopic c-*fos*, these neuroectoderm-derived cells were able to differentiate into cytologically normal structures.

Analysis of c-*fos* expression in the limbs and bones of nontransgenic and transgenic animals revealed the following: Osteoblasts rather than osteoclasts (Sandberg et al. 1988) were found to express c-*fos* in the growth plates of developing long bones. The transgenic mice express high levels of c-*fos* in the epiphysial cartilage, in the region of perichondrial ossification, and within the tendons as early as embryonic stage E17.5. Thus, the onset of the observed phenotype (hyperplasia 2–3 weeks after birth) is preceded by exogenous c-*fos* expression and can be attributed to elevated levels of c-*fos* in osteoblasts in the region of perichondrial ossification and to chondroblasts in the epiphyseal cartilage. No significant differences of c-*fos* expression could be found within the regions of endochondrial ossification of normal and transgenic animals at any stage of the investigations.

Within the tumors themselves, the majority of cells that express exogenous c-*fos* are still unidentified. However, osteoblasts and chondroblasts, which express c-*fos*, could be detected in newly formed cartilage and bone tissue in these tumors. This further emphasizes the role of osteoblasts and chondroblasts in the development of the observed bone hyperplasia and bone tumors. It is interesting to note that all the cell types that show ectopic c-*fos* expression at E17.5 and P2 seem to be present in the bone tumors.

As mentioned above, our experimental approach, using a 1-kb fragment of the murine c-*fos* cDNA as a probe, does not allow us to distinguish between endogenous c-*fos* and the exogenous transgene. Experiments using a specific probe such as an MT-specific oligo are in progress to substantiate our results. We are also in the process of using immunocytochemistry to define better the cell types expressing c-*fos* during bone formation in development and in the bone remodeling process in adult animals. Currently, we are analyzing primary cells derived from affected and unaffected bone tissue to possibly identify the *fos*/AP-1 responsive genes, as well as the genes leading to tumor formation. In parallel, we are generating new transgenic mouse lines using mutant c-*fos* genes, as well as homologous recombination in ES cells to inactivate the c-*fos* gene. All these projects aim at a better definition of the molecular mechanisms of *fos* action and its role in growth control, differentiation, and in tumorigenesis.

ACKNOWLEDGMENTS

We thank Drs. Carola Dony and Gregory Dressler for help in setting up the in situ hybridization technique, Adriano Aguzzi and Otmar Wiestler for their enormous help in identifying cell types of the nervous system, Michael Veit Krenn, Ulrich

Rüther, and our colleagues at the European Molecular Biology Laboratory and the Research Institute of Molecular Pathology for discussions and continuous interest in this project. K.H. was funded by a long-term Erwin Schrödinger Fellowship from the Bundesministerium für Wissenschaft und Forschung in Austria.

REFERENCES

Caubet, J.F. 1989. c-fos proto-oncogene expression in the nervous system during mouse development. Mol. Cell. Biol. 9: 2269.

Chiu, R., W.J. Boyle, J. Meek, T. Smeal, T. Hunter, and M. Karin. 1988. The fos protein interacts with c-jun/AP-1 to stimulate transcription of AP-1 responsive genes. Cell 54: 541.

Curran, T. and B.R. Franza, Jr. 1988. fos and jun: The AP-1 connection. Cell 55: 395.

De Togni, P., H. Niman, V. Raymond, P. Sawchenko, and I.M. Verma. 1988. Detection of fos protein during osteogenesis by monoclonal antibodies. Mol. Cell. Biol. 8: 2251.

Dony, C. and P. Gruss. 1988. Proto-oncogene c-fos expression in growth regions of fetal bone and mesodermal web tissue. Nature 328: 711.

Gonda, T.J. and D. Metcalf. 1984. Expression of myb, myc and fos proto oncogenes during the differentiation of a murine myeloid leukaemia. Nature 310: 249.

Holt, I.T., T. Venkat Gopal, A.D. Moulton, and A.W. Nienhuis. 1986. Inducible production of c-fos antisense RNA inhibits 3T3 cell proliferation. Proc. Natl. Acad. Sci. 83: 4794.

Kreipe, H., H.J. Radzun, K. Heidorn, C. Mader, and M.R. Parwaresh. 1987. Human neutrophilic and eosinophilic granulocytes display different levels of c-fos proto-oncogene expression: An in situ hybridization study. J. Histochem. Cytochem. 35: 837.

Müller, R. 1986. Cellular and viral fos genes: Structure, regulation of expression and biological properties of their encoded products. Biochim. Biophys. Acta 823: 207.

Nishikura, K. and J.M. Murray. 1987. Antisense RNA of proto-oncogene c-fos blocks renewed growth of quiescent 3T3 cells. Mol. Cell. Biol. 7: 639.

Rauscher, F.J., III, D.R. Cohen, T. Curran, T.J. Bos, P.K. Vogt, D. Bohmann, R. Tjian, and B.R. Franza, Jr. 1988. Fos associated protein p39 is the product of the jun proto oncogene. Science 240: 1010.

Rüther, U., E.F. Wagner, and R. Müller. 1985. Analysis of the differentiation-promoting potential of the inducible c-fos genes introduced into embryonal carcinoma cells. EMBO J. 4: 1775.

Rüther, U., D. Komitowski, F.R. Schubert, and E.F. Wagner. 1989. c-fos expression induces bone tumors in transgenic mice. Oncogene 4: (in press).

Rüther, U., C. Garber, D. Komitowski, R. Müller, and E.F. Wagner. 1987. Deregulated c-fos expression interferes with normal bone development in transgenic mice. Nature 325: 412.

Rüther, U., W. Müller, T. Sumida, T. Tokuhisa, K. Rajewsky, and E.F. Wagner. 1988. c-*fos* expression interferes with thymus development in transgenic mice. *Cell* **53:** 847.

Sandberg, M., T. Vuorio, H. Hirvonen, K. Alitalo, and E. Vuorio. 1988. Enhanced expression of TGF-β and c-*fos* mRNAs in the growth plates of developing human long bones. *Development* **12:** 461.

Sassone-Corsi, P., W.W. Lamph, M. Kamps, and I.M. Verma. 1988. *Fos* associated cellular p39 is related to nuclear transcription factor AP-1. *Cell* **54:** 553.

Verma, I.M. and P. Sassone-Corsi. 1987. Proto-oncogene *fos*: Complex but versatile regulation. *Cell* **51:** 513.

Wagner, E.F., R.L. Williams, and U. Rüther. 1989. c-*fos* and *polyoma middle* T oncogene expression in transgenic mice and embryonal stem cell chimaeras. In *Cell to cell signals in mammalian development* (ed. S.W. de Laat et al.), p. 301. Springer-Verlag, Berlin.

int-1 and Pattern Regulation

A.P. McMahon[1] and R.T. Moon[2]

[1]Department of Cell and Developmental Biology, Roche Institute of Molecular Biology, Roche Research Center, Nutley, New Jersey 07110

[2]Department of Pharmacology, School of Medicine, University of Washington, Seattle, Washington 98195

int-1 is one of a number of cellular oncogenes activated by proviral insertion (Nusse 1986). Insertion of mouse mammary tumor virus 5′ or 3′ of the *int-1* locus but outside of the coding sequence (van Ooyen and Nusse 1984) drives inappropriate expression of the *int-1* gene in the mammary gland. In certain strains of mice and when this situation is recreated in transgenic mice (Tsukamoto et al. 1988), mammary tumors arise at high frequency. Recent attention has focused on the normal role of the *int-1* protein, which lies in the development of distantly related species such as *Drosophila* and mice. In *Drosophila*, *d-int-1* is involved in the normal determination of segmental pattern and encodes the gene *wingless* (*wg*) (Rijsewijk et al. 1987). *d-int-1* is required in all segments. Absence of the protein results in the loss of posterior segmental pattern elements, whereas anterior structures, in particular the denticle bands, are duplicated in mirror image symmetry and come to fill most of the mutant segment. Thus, *d-int-1* falls into a class of genes, the segment polarity genes (Nüsslein-Volhard and Wieschaus 1980), that regulate patterning events. Unlike other segment polarity mutants, small clones of *wg* mutant cells can develop normally. Thus, mutant cells are rescued by neighboring wild-type cells (Morata and Lawrence 1977). This result and analysis of the predicted *d-int-1* protein sequence, which contains a signal peptide sequence, suggest that *int-1* is a secreted protein.

int-1 Expression in Mouse Development

We have examined expression of *int-1* in mouse development by in situ hybridization (Wilkinson et al. 1987). *m-int-1* RNA is not detected prior to neural plate formation, but on neurulation *int-1* transcripts accumulate specifically in the developing neural regions. Expression outside of the neural regions was not detected at any fetal stage of development. At early somite stages (8.5-days postcoitus), expression was restricted to a relatively broad region of the rostral neural plate in presumptive brain regions. No caudal expression in presumptive spinal cord regions was detected at this time. As the neural plate folds and fuses to form the neural tube, *int-1* expression becomes localized to the dorsal midline from the midbrain to the caudal extremity of the neural tube. In the developing mid- and hindbrains, some *int-1* expression does occur in other regions. However, the predominate expression is at the dorsal midline or the equivalent cells that are laterally displaced in the hindbrain. The basic pattern of *int-1* expression, which is established by 10.5 days of development, changes little over the next 4 days, a period in which the neural tube undergoes considerable morphological and functional specialization. Figure 1 illustrates the highly localized expression of *int-1* in the developing spinal cord at 10.5 days of development. Interestingly, expression occurs symmetrically about the dorsal ventral axis of symmetry, which splits the neural tube into left and right halves (see Fig. 1). Thus, *int-1* expression is restricted to a region in which pattern-regulating molecules may be localized.

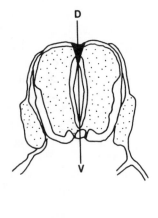

Figure 1 *int-1* expression in the mouse neural tube: (*a*) In situ hybridization showing *int-1* expression in the roof plate at the dorsal midline of the neural tube; and (*b*) schematic representation of *int-1* expression (shaded) at the dorsal ventral axis of symmetry of the spinal cord. (Reprinted, with permission, from McMahon and Moon 1989.)

Ectopic Expression of *int-1* Causes Pattern Abnormalities

To examine whether *int-1* is involved in pattern regulation in vertebrates, we have attempted to interfere with normal *int-1* expression in vivo. As a model system, we have studied the effects of ectopic expression of *m-int-1* in *Xenopus* development. The *Xenopus int-1* gene shares considerable sequence identity with its mouse homolog ([69%] Nordermeer et al. 1989). Although the spatial localization and the exact onset of *Xenopus int-1* expression is not known, no *int-1* RNA is detected until early neurula stages. Fertilized eggs were injected with approximately 2 ng of polyadenylated mouse *int-1* RNA in either animal or vegetal poles with similar results. Embryos developed to gastrulae normally, but by neurula stages the neural plate was clearly abnormal. The anterior neural plate was bifurcated into two distinct axes, whereas posteriorly the neural plate was enlarged (Fig. 2). Histological examination indicated that the underlying cause of the neural plate abnormalities was a duplication of the axial notochord, which is responsible for neural induction. In posterior regions, the notochord was fused into a single enlarged notochord that split anteriorly into two distinct branches of equivalent size to the single notochord in control embryos. As well as a duplication of the axial structures, the paraxial mesoderm (somites) was also duplicated. Therefore, much of the normal anterior-posterior pattern elements were duplicated in these experiments.

Examination of the translation of injected RNA indicated that *int-1* protein is widely dispersed. Thus, a specific and rath-

Figure 2 Bifurcated *Xenopus* neurula following *int-1* RNA injection. (*Left*) Bifurcated embryo injected with wild-type *int-1* RNA. (*Right*) Control embryo. Arrows indicate the anterior neural tube. (Reprinted, with permission, from McMahon and Moon 1989.)

er striking alteration in axial specification is produced by a rather crude misexpression of *int-1* protein. This effect is dependent on producing normal *int-1* protein, since this phenotype was not obtained when RNAs encoding mutated forms of the *int-1* protein were injected. Moreover, the observed effects are highly reproducible. Over 90% of surviving neurulae injected with wild-type *int-1* RNA develop a bifurcated anterior neural tube.

These results clearly demonstrate that, as in *Drosophila*, altered expression of *int-1* can affect patterning events in vertebrates. Duplication of the embryonic axis presumably results from a perturbation of the normal axial specification system, which operates during gastrulation. However, in light of what is known of mouse *int-1* expression, we would not have expected a role for *int-1* in developmental events preceding neural tube formation. A thorough examination of *int-1* expression in *Xenopus* and mice, together with appropriate experimental manipulation as reported here, will be required for a real understanding of the roles of *int-1* in vertebrate development.

REFERENCES

McMahon, A.P. and R.T. Moon. 1989. *int-1*—A proto-oncogene involved in cell signalling. *Development* (in press).

Morata, G. and P.A. Lawrence. 1977. The development of *wingless,* a homeotic mutation of *Drosophila. Dev. Biol.* **56:** 227.

Nordermeer, J., F. Meijlink, P. Verrijzer, F. Rijsewijk, and O. Destrée. 1989. Isolation of the *Xenopus* homologue of *int-1 wingless* and expression during neurula stages of early development. *Nucleic Acids Res.* **17:** 11.

Nusse, R. 1986. The activation of cellular oncogenes by retroviral insertion. *Trends in Genet.* **2:** 244.

Nüsslein-Volhard, C. and E. Wieschaus. 1980. Mutations affecting segment number and polarity in *Drosophila. Nature* **287:** 795.

Rijsewijk, F., M. Schuermann, E. Wagenaar, P. Parren, D. Weigel, and R. Nusse. 1987. The *Drosophila* homolog of the mouse mammary oncogene *int-1* is identical to the segment polarity gene *wingless. Cell* **50:** 649.

Tsukamoto, A.S., R. Grosschedl, R.C. Guzman, T. Parslow, and H.E. Varmus. 1988. Expression of the *int-1* gene in transgenic mice is associated with mammary gland hyperplasia and adenocarcinomas in male and female mice. *Cell* **55:** 619.

van Ooyen, A. and R. Nusse. 1984. Structure and nucleotide sequence of the putative mammary oncogene *int-1*; proviral insertions leave the protein coding domain intact. *Cell* **39:** 233.

Wilkinson, D.G., J.A. Bailes, and A.P. McMahon. 1987. Expression of the proto-oncogene *int-1* is restricted to specific neural cells in the developing mouse embryo. *Cell* **50:** 79.

int-1 and *int-4*, Two Genes Active in Mouse Mammary Tumorigenesis and in Normal Embryogenesis

R. Nusse, M. van den Heuvel, H. Roelink,
J. Knol, and M. van de Vijver

Division of Molecular Biology, Netherlands Cancer Institute
1066 CX Amsterdam, The Netherlands

Our laboratory investigates the molecular mechanism of viral mammary tumorigenesis in mice. The virus responsible for the tumors, the mouse mammary tumor virus is a retrovirus that, in the course of infection, inserts proviral DNA in the genome of the target cell. Insertion is a mutagenic event, leading to the reproducible transcriptional activation of a small set of genes, collectively called *int*. Through provirus tagging methods, four different *int* genes have been identified over the past few years (Table 1). These genes are not structurally related to each other. Surprisingly, at least three *int* genes appear to share, besides their implication in mammary cancer, the property of being expressed in a very restricted pattern in embryogenesis (for review, see Nusse 1988a,b). In this short paper, our recent work on the mouse *int-1* and *int-4* genes and the *Drosophila int-1/wingless* gene will be discussed.

The *int-1* Gene in Mammary Tumors and in Normal Development

The *int-1* gene can cause cell transformation in culture and can lead to mammary cancer as a transgene (Tsukamoto et al. 1988). The gene encodes a cysteine-rich protein, with a typical signal peptide (van Ooyen and Nusse 1984). During its synthesis and transport, the *int-1* protein is glycosylated and enters the secretory pathway (Papkoff et al. 1987), suggesting that the protein is released from cells. The structure of the *int-1* protein is highly conserved between species, including humans and *Drosophila* (van Ooyen et al. 1985, Rijsewijk et al. 1987).

Table 1 Properties of the *int* Genes

| Gene | Chromosome | | Related to [a] | Protein |
	mouse	human		
int-1	15	12	*irp, wingless*	41 kD, signal peptide
int-2	7	11	FGF, *hst*	27 kD, signal peptide
int-3	17	?	?	?
int-4	11	17	?	?

[a](*irp*) *int*-1-related protein and (FGF) fibroblast growth factor.

In mouse embryos, the *int-1* gene is transiently expressed in areas of the developing neural system (Wilkinson et al. 1987). Retinoic-acid-treated P19 embryonal carcinoma cells have often been used as an in vitro model for the molecular basis of neural development. The P19 cell line is a pluripotent embryonal carcinoma line derived from a 7.5-day-old embryo. The cells can differentiate into multiple pathways, depending on the nature and the dose of the inducing agent. High (10^{-8}–10^{-7} M) doses of retinoic acid elicit differentiation into neuron-like cells, marked by the appearance of neural cell proteins. We recently found that *int-1* is transiently expressed in differentiating P19 cells (Schuuring et al. 1989). The time course and retinoic acid dose dependence of *int-1* expression suggest that the gene is specifically expressed during early neural differentiation. P19 cells may be a useful model to study, at the cellular level, the role of *int-1* in neural development.

We have started to locate the *int-1* promoter region responsible for the retinoic-acid-mediated induction of transcription. By primer extensions and nuclease protection experiments, we have shown that there are two *int-1* promoters 160 nucleotides from each other. One start of transcription is preceded by a TATA box; the more upstream start is preceded by a GC box. Constructs were made in which the promoter of *int-1* was fused to the bacterial chloramphenicol acetyltransferase gene. Transfection of these constructs into P19 cells showed that retinoic-acid-inducibility was conferred by a 1.5-kb upstream region.

Much of our effort to understand the mechanisms of action of *int-1* is concentrated on the properties of the protein. In collaboration with H. Stunnenberg, (EMBL, Heidelberg) recombinant vaccinia viruses have been made containing the mouse and the *Drosophila int-1* genes. On infection of various cell lines and labeling with radioactive amino acids, the *int-1* protein can be detected in a whole-cell extract as prominent bands on autoradiograms of SDS gels. Antibodies to synthetic peptides or to bacterial fusion proteins recognize the vaccinia-made *int-1* proteins on Western blot analysis. We intend to use this expression system to purify *int-1* protein in a biologically active form.

int-1 in *Drosophila* Is Identical to *wingless*

int-1 is the mouse homolog of the *Drosophila* segment polarity gene *wingless* (Baker 1987; Rijsewijk et al. 1987). The non-autonomous character of *wingless* mutant cells (Wieschaus and Riggleman 1987) indicates that the *wingless* gene product is secreted and determines the fate of surrounding cells, a conclusion corroborated by its homology with *int-1*. In *Drosophila*, a large array of mutants exists, all characterized to be deficient in embryonal development (Nüsslein-Volhard and Wieschaus 1980). Many of the genes responsible for the mutations affecting segment number and polarity have now been cloned, and are thought to interact with each other to generate the basic body plan of the fly. A major goal of our research is to characterize the *wingless* protein and to define interactions of *wingless* with other gene products, both at the genetic and at the protein level.

With a bacterial expression system, we made fusion proteins containing *wingless* antigens. Immunostaining of whole mount *Drosophila* embryos with the purified antisera to the bacterial fusion proteins, performed in collaboration with P. Lawrence (MRC, Cambridge), showed specific localization of the *wingless* protein (M. van den Heuvel et al., in prep.). Staining is found in different regions of the head (foregut primordium and presumably the antennal region and the labral region), in the hindgut/anal region, and in 15 stripes in the trunk region (Fig. 1). No staining was found in *wingless* mutant embryos. The protein is detected in approximately the same regions of the embryo as the *wingless* mRNA (Baker 1987).

The segment polarity genes *engrailed* and *wingless* are thought to be expressed on either site of the parasegment border (Baker 1987). From genetic experiments, it is known that the *wingless* gene product is necessary for the maintenance of *engrailed* expression (DiNardo et al. 1988). In a double-staining of wild-type embryos with both the *wingless* and the *engrailed* antibodies, we find the proteins localized in the same cells as defined by the in situ hybridization. The *engrailed* protein is located in the nucleus of the anterior-most cells of every para-

Figure 1 A *Drosophila* embryo (extended germ-band stage) stained with an antibody to the *wingless* protein. *wingless* is a segment polarity gene that is necessary for correct differentiation within each segment. The mouse homolog of *wingless*, *int-1*, is involved in the development of the nervous system and can behave as an oncogene in mouse mammary tumors. For details, see Nusse et al. and McMahon and Moon (both this volume). (Photograph courtesy of M. van den Heuvel and R. Nusse, and P. Lawrence, Medical Research Council, Cambridge, England.)

segment; the *wingless* protein seems to be located on the cell membrane of the posterior-most cells of every parasegment.

In the electron microscope, *wingless* protein is seen in small membrane-bound vesicles, in multivesicular bodies, and in the intercellular space. The multivesicular bodies containing the *wingless* protein are occasionally found in *engrailed*-positive cells, suggesting that the *wingless* protein behaves as a paracrine signal.

int-4 Proto-oncogene

This gene is infrequently (5%) activated in mammary tumors induced by the GR strain of mouse mamary tumor virus. With the aid of N. Copeland (NCI, Frederick), using a series of recombinant inbred mouse strains, it was established that *int-4* is located on the tip of chromosome 11, 4.2-cM distal from a cluster of genes including c-*erbB2* and c-*erbA*. This location excludes any identity with other known genes.

Two differently sized *int-4* mRNAs of 2.1 kb and 3.7 kb are found in tumors and in P19 cells. We employed a sensitive detection system, using antisense RNA probes in an RNase protection assay, to search for normal expression of *int-4*. These assays showed that *int-4* is expressed during embryogenesis, starting at day 10 of gestation. At the end of gestation, *int-4* expression is still detectable. A section of a 16-day-old embryo showed that *int-4* expression is located in the dorsal side and the head. In an adult mouse, we found *int-4* expression in the central nervous system only, both in the forebrain and in the hindbrain. (H. Roelink et al., in prep.).

Like *int-1*, *int-4* is also expressed in retinoic-acid-treated P19 embryonic carcinoma cells. *Int-4* RNA becomes detectable after 6 days of culturing the cells as aggregates, reaches peak levels after 2 days of culturing, and has disappeared after 2 weeks. The *int-4* gene is not expressed in F9 cells induced with retinoic acid. These expression data suggest a role of *int-4* in the developing and adult nervous system.

In conclusion, it appears that genes active in mouse mammary cancer have a normal function in early developmental decisions. Besides the genes discussed here, a function in embryogenesis has also been proposed for *int-2*, a member of the fibroblast growth factor family (Dickson and Peters 1987). *int-2* is expressed in many embryonal carcinoma cells, and its spatial pattern of expression in mouse embryos indicates that the gene may have a role in many steps in development, as early as during migration of mesodermal cells through the primitive streak (Wilkinson et al. 1988).

One wonders whether there is a common mechanism of action of the *int* genes, either in tumorigenesis or in embryogenesis, despite the lack of sequence relationship. It may be that developmental genes are potent mammary oncogenes because the tumors arise in a developing organ—only in mammary glands cycling through multiple rounds of pregnancy. The mitogenic stimulus comes then from pregnancy hormones, and aberrant growth may result from activation of genes controlling pattern formation.

REFERENCES

Baker, N.E. 1987. Molecular cloning of sequences from *wingless*, a segment polarity gene in *Drosophila*: The spatial distribution of a transcript in embryos. *EMBO J.* **6**: 1765.

Dickson, C. and G. Peters. 1987. Potential oncogene product related to growth factors. *Nature* **326**: 833.

DiNardo, S., E. Sher, J. Heemskerk-Jongens, J.A. Kassis, and P.H. O'Farrell. 1988. Two-tiered regulation of spatially patterned *engrailed* gene expression during *Drosophila* embryogenesis. *Nature* **332**: 604.

Nusse, R. 1988a. The activation of cellular oncogenes by proviral insertion in murine mammary tumors. *In Breast cancer: Cellular and molecular biology* (ed. M.E. Lippman and R. Dickson) p. 283. Kluwer Academic Publishers, Boston.

———. 1988b. The *int* genes in mammary tumorigenesis and in normal development. *Trends Genet.* **4**: 291.

Nüsslein-Volhard, C. and E. Wieschaus. 1980. Mutations affecting segment number and polarity in *Drosophila*. *Nature* **287**: 795.

Papkoff, J., A.M.C. Brown, and H.E. Varmus. 1987. The *int-1* proto-oncogene products are glycoproteins that appear to enter the secretory pathway. *Mol. Cell. Biol.* **7**: 3978.

Rijsewijk, F., M. Schuermann, E. Wagenaar, P. Parren, D. Weigel, and R. Nusse. 1987. The *Drosophila* homologue of the mammary oncogene *int-1* is identical to the segment polarity gene *wingless*. *Cell* **50**: 649.

Schuuring, E., E. van Deemter, H. Roelink, and R. Nusse. 1989. Expression of the *int-1* proto-oncogene during differentiation of P19 embryonal carcinoma cells. *Mol. Cell. Biol.* **9**: 1357.

Tsukamoto, A.S., R. Grosschedl, R.C. Guzman, T. Parslow, and H.E. Varmus. 1988. Expression of the *int-1* gene in transgenic mice is associated with mammary gland hyperplasia and adenocarcinomas in male and female mice. *Cell* **55**: 619.

van Ooyen, A. and R. Nusse. 1984. Structure and nucleotide sequence of the putative mammary oncogene *int-1*: Proviral insertions leave the protein-encoding domain intact. *Cell* **39**: 233.

van Ooyen, A., V. Kwee, and R. Nusse. 1985. The nucleotide sequence of the human *int-1* mammary oncogene; evolutionary conservation of coding and noncoding sequences. *EMBO J.* **4**: 2905.

Wieschaus, E. and R. Riggleman. 1987. Autonomous requirements for the segment polarity gene *armadillo* during *Drosophila* embryogenesis. *Cell* **49:** 177.

Wilkinson, D.G., J.A. Bailes, and A.P. McMahon. 1987. Expression of the proto-oncogene *int-1* is restricted to specific neural cells in the developing mouse embryo. *Cell* **50:** 79.

Wilkinson, D.G., G. Peters, C. Dickson, and A.P. McMahon. 1988. Expression of the FGF-related proto-oncogene *int-2* during gastrulation and neurulation in the mouse. *EMBO J.* **7:** 691.